T0271833

Fundamentals of Offshore Engineering

Fundamentals of Offshore Engineering addresses the basics of design for offshore oil and gas production systems and examines the health, safety, and environmental (HSE) aspects in the oil and gas industry with emphasis toward safety measures in design and operations. It also covers fundamental issues of crude oil and natural gas exploration and extraction and also includes coverage of seismic surveys and green energy systems. Details of offshore platforms, describing the types, historical development, basics of analysis and design, environmental loads, and potential hazards are also provided. The book serves as a useful resource for universities that teach offshore engineering to senior undergraduate and graduate students as well as a guide for practicing engineers.

- Includes coverage of wave loads, wind loads, ice loads, and fire loads on structures.
- Discusses offshore pipelines and subsea engineering to help readers understand the fundamentals of petroleum production and related pipeline installation.

Fundamentals of Offshore Engineering

Srinivasan Chandrasekaran,
Surasak Phoemsapthawee,
Shanker Krishna, and Hari Sreenivasan

CRC Press
Taylor & Francis Group
Boca Raton London New York

CRC Press is an imprint of the
Taylor & Francis Group, an **informa** business

Designed cover image: © Shutterstock

First edition published 2025
by CRC Press
2385 NW Executive Center Drive, Suite 320, Boca Raton FL 33431

and by CRC Press
4 Park Square, Milton Park, Abingdon, Oxon, OX14 4RN

CRC Press is an imprint of Taylor & Francis Group, LLC

© 2025 Srinivasan Chandrasekaran, Surasak Phoemsapthawee,
Shanker Krishna, and Hari Sreenivasan

Reasonable efforts have been made to publish reliable data and information, but the author and publisher cannot assume responsibility for the validity of all materials or the consequences of their use. The authors and publishers have attempted to trace the copyright holders of all material reproduced in this publication and apologize to copyright holders if permission to publish in this form has not been obtained. If any copyright material has not been acknowledged please write and let us know so we may rectify in any future reprint.

Except as permitted under U.S. Copyright Law, no part of this book may be reprinted, reproduced, transmitted, or utilized in any form by any electronic, mechanical, or other means, now known or hereafter invented, including photocopying, microfilming, and recording, or in any information storage or retrieval system, without written permission from the publishers.

For permission to photocopy or use material electronically from this work, access www.copyright.com or contact the Copyright Clearance Center, Inc. (CCC), 222 Rosewood Drive, Danvers, MA 01923, 978-750-8400. For works that are not available on CCC please contact mpkbookspermissions@tandf.co.uk

Trademark notice: Product or corporate names may be trademarks or registered trademarks and are used only for identification and explanation without intent to infringe.

ISBN: 978-1-032-80606-8 (hbk)
ISBN: 978-1-032-80605-1 (pbk)
ISBN: 978-1-003-49766-0 (ebk)

DOI: 10.1201/9781003497660

Typeset in Times LT Std
by Apex CoVantage, LLC

Contents

Preface

Fundamentals of Offshore Engineering attempts to meet the classroom learning requirements of this complex subject through simple illustrations and discussions. Various environmental issues related to offshore oil and gas exploration and their consequences are presented in detail. A detailed numerical model of atmospheric pollution assessment, presented in the book, will be helpful for the learners to compute the pollution concentrations arising from oil and gas exploration, production, and processing. The authors have described a few accident case studies, based on their industrial experience, which highlight the major causes and consequences of environmental pollution. The book also discusses HSE practices that are commonly followed in the process industries. Details of risk assessment methods and procedures discussed by the authors will be helpful to the readers. An interesting tool for logical risk analysis helps to perform risk assessment and mitigation, which is also illustrated with an example. Hazop methodology, useful for detecting potential hazards and operability problems in process industries, will serve as a useful tool for practicing engineers.

A complete chapter on the origin of crude oil and natural gas with their geophysical properties adds value to the contents of the book and enhances wider readability. A brief discussion on non-conventional energy sources and green energy systems provides an overview and their relevance in the present context. A chapter on offshore platforms deals with their historical development and their structural action under different environmental loads. The newly evolved structural forms and their discrete characteristics are also discussed. This chapter helps the reader understand the structural action of different forms in the offshore platform, leading to the need to upgrade structural systems with time. An overview of the construction stages of offshore plants and their foundation systems helps understand the construction methods and the allied complexities.

Offshore drilling and production, dealing with the mechanics involved in petroleum production, is explained through illustration of types of drilling rigs and their working principles. A short description of blowouts and kicks helps understand the reasons for their occurrence and methods to avoid or control their risk occurrence. A brief section on oil recovery helps understand the basics of EOR and thermal recovery methods. A brief note on oil spills, consequences, and prevention helps readers learn this important segment of environmental pollution arising from oil and gas exploration. A brief chapter on subsea systems illustrates their vital components and layout. It briefs on the location importance and nature of their function in the subsea system. Production trees with the key role of jumpers and umbilical are also discussed in detail.

The authors are convinced by a large brace of the contents, covering offshore structural engineering, loads and actions, petroleum geology, and production methods. This book will be an ideal and primary reference for engineers of various disciplines; namely, civil, mechanical, offshore, petroleum, chemical, naval architecture, and safety executives. A rich teaching experience commended by the authors and a sound knowledge base, established through their publications in refereed journal articles, are testimonies to confirm the previously mentioned declaration.

The authors sincerely thank the chairman, Center for Outreach and Digital Education, Indian Institute of Technology, Madras, for the administrative support extended to publish this book. The authors also extend thanks for the support and cooperation extended by the Faculty of International Maritime Studies, Kasetsart University, Sriracha campus, Thailand. The lead author, in the capacity of Visiting Professor, thanks the director, International College, Kasetsart University, Bangkok campus, for the administrative support and cooperation extended to publish this book. All authors sincerely thank their colleagues, research scholars, administrative authorities, and researchers cited in the book for their premier contributions to the subject.

The authors also thank the director general and the president of the Pandit Deendayal Energy University for the administrative support extended in publishing this book.

Srinivasan Chandrasekaran
Dept. of Ocean Engg, IIT Madras, India
Faculty of International Maritime Studies,
Kasetsart University, Sriracha campus, Thailand
International College, Kasetsart University
Bangkok, Thailand

Surasak Phoemsapthawee
Dept. of Maritime Engineering,
Faculty of International Maritime Studies
Kasetsart University, Sriracha campus, Thailand

Shanker Krishna
Department of Petroleum Engineering,
School of Energy Technology
Pandit Deendayal Energy University (PDEU),
Gandhinagar, Gujarat

Hari Sreenivasan
Department of Petroleum Engineering,
School of Energy Technology
Pandit Deendayal Energy University (PDEU),
Gandhinagar, Gujarat

About the Authors

Srinivasan Chandrasekaran is currently a professor with a Higher Academic Grade at the Department of Ocean Engineering, Indian Institute of Technology, Madras. He is also a visiting professor at the Faculty of International Maritime Studies, Kasetsart University, Sriracha campus, Thailand, and a visiting professor at International College, Kasetsart University, Bangkok campus, Thailand. He has a rich experience in teaching, research, and industrial consultancy of about 29 years. He has supervised many sponsored research projects and offshore consultancy assignments, both in India and abroad. His active research areas include dynamic analysis and design of offshore structures, structural health monitoring of ocean structures, risk and reliability, fire resistant design of structures, use of functionally graded materials (FGM) in marine risers, and health, safety and environmental (HSE) management in process industries. He was a visiting fellow under the Ministry of Italian University Research (MiUR) invitation to the University of Naples Federico II for two years. During his recent position as a visiting professor in Italy, he participated in ongoing research activities concerning response control of structures and published a few papers with international cooperation. He has authored about 180 research papers in peer-reviewed international journals and refereed conferences organized by professional societies worldwide. He has authored 22 textbooks, which various publishers of international repute publish. He is an active member of several professional bodies and societies, both in India and abroad. He has also conducted about 20 distance-education programs on various engineering subjects for the National Program on Technology-Enhanced Learning (NPTEL), Govt. of India. He is a vibrant speaker and has delivered many keynote addresses at international conferences, workshops, and seminars organized in India and abroad. He is designated as a top researcher, listed within 2% of experts in the world, as quoted by an independent survey conducted by Google and Stanford University, USA.

Surasak Phoemsapthawee Dr. Surasak Phoemsapthawee is an assistant professor specializing in naval architecture and offshore structures in the Department of Maritime Engineering, Faculty of International Maritime Studies, Kasetsart University, Sri Racha campus, Thailand. His primary research interests include hydrodynamic simulation of gliders for coastal and multi-sensors application, sheet cavitation flow in unsteady potential-flow simulation over hydrofoil and propeller, and the design of offshore structures. He is the nodal faculty involved in training graduates for offshore structural engineering and has wide international exposure with École Nationale Supérieure de Techniques Avancées de Bretagne (ENSTA-Bretagne), France, Université Européenne de Bretagne, France, and Service Hydrodynamique et Océano-météo, Ifremer, France. He is a member of many international professional societies and has published research papers in leading Scopus-indexed journals in the domain of naval architecture, ocean engineering, ship technology and research, and others.

Shanker Krishna Dr. Shanker Krishna is an assistant professor in the Department of Petroleum Engineering, School of Energy Technology, at Pandit Deendayal Energy University (PDEU). He is the charge of the Drilling, Cementing, and Stimulation Research Centre at PDEU. His areas of expertise are geomechanics and petrophysics, hydraulic fracturing, enhanced oil recovery, gas hydrates, rock mechanics, and material characterization. He has published many research papers in refereed journals and presented his works in international conferences of high repute. He is also an active member of many professional societies in India and abroad.

Hari Sreenivasan Dr. Hari Sreenivasan is an assistant professor in the Department of Petroleum Engineering, School of Energy Technology, at Pandit Deendayal Energy University (PDEU). He is the charge of the Petroleum Engineering Lab at PDEU. His areas of expertise are offshore engineering, functionally graded materials, drilling and production of oil and gas engineering. He has published many research papers in refereed journals and presented his works at international conferences of high repute. He is also an active member of many professional societies in India and abroad. He has also been granted a patent on the design and product development of FGM for corrosion resistance applications.

Figures

CHAPTER 3: CRUDE OIL AND NATURAL GAS

CHAPTER 4: OFFSHORE PLATFORMS

CHAPTER 5: PETROLEUM PRODUCTION

CHAPTER 6: SUBSEA SYSTEMS

1 Offshore Oil and Gas Industry

Summary

This chapter presents major environmental issues related to offshore oil and gas exploration and their consequences. An overview of oil spills, drilling discharges into the ocean, and measures to reduce the consequences are discussed. Environmental protection regarding ecological monitoring and management standards are also discussed briefly. Detailed numerical models to assess atmospheric pollution arising from oil and gas processing plants are helpful for the learners to compute the pollution concentrations. A few accident case studies, highlighting the causes and consequences of environmental pollution, are presented.

1.1 ENVIRONMENTAL ISSUES AND MANAGEMENT

The primary environmental issues are the considerable impact caused by oil and gas production on the shelf ecosystems and marine biological resources. It contributes to the disturbance of the life hierarchy at different levels and significantly influences the aquatic ecosystem. Most importantly, it is to be noted that the biological consequences of accidental oil spills into the marine environment are irreversible. Environmental pollution in marine ecosystems creates complexities and stands as a visible consequence for various emerging problems in environmental management. It results in an uneven distribution of marine life and its concentration in the shelf and coastal zone, the habitat for about 90% of the marine commercial organisms (Scheidegger, 1963; Schwartz, 2005; Skelton, 1997). Most known oil and gas fields are located in this zone, causing severe ecological disturbances. Crude oil and natural gas play a significant and crucial role in contributing to the total energy produced in the world. It is still increasing due to the high demand and increased energy consumption. The historical development is remarkable for its high dynamics, rapid technological progress, vast geography of exploration, and comprehensive production activities. Table 1.1 summarizes various energy resources. It is seen from the table that there has been significant growth and relative stabilization in the recent past; the decrease in oil production in large regions is also significantly noticeable (Cairns, 1992; Vinnem, 2007).

Hydrocarbon exploring fields located inland are depleted, and the focus is shifted toward the shelf resources. This shift to the continental shelf is foreseen to significantly affect marine organisms' growth (Patin, 1999). It is also known that improvements in drilling technologies led to exploration possibilities in the polar region. Advanced technology and the latest equipment for developing offshore hydrate resources pose a serious environmental threat. Many mechanical and chemical

DOI: 10.1201/9781003497660-1

techniques used for oil exploration, production enhancement, and processing cause severe environmental issues. For example, hot water pumping and the introduction of inhibitors like methanol have posed severe challenges to the marine environment. The continental shelf, the main arena for shipping and fishing, is now being explored for oil and gas. Prospective locations of oil and gas fields in the shelf zones often overlap with the regions of high biological productivity. Recent explorations of gas hydrates, which are highly promising, are found in marine areas; their development is envisaged as a potential threat to the marine environment.

1.2   ANTHROPOGENIC IMPACT OF HYDROSPHERE

"Anthropogenic impact" refers to assessing the state of hydrosphere and water eco-systems. It depends on many criteria such as changes in temperature regime, radioactive background, discharges of toxic effluents, inflow of nutrients, irretrievable water consumption, damage of water organisms during seismic surveys, landing of commercial species and their cultivation, destruction of the shoreline, etc. The anthropogenic impact on the hydrosphere by offshore oil and gas production and land-based activities are given in Tables 1.2 and 1.3, respectively. Anthropogenic impact on marine and freshwater systems causes hidden disturbances to water communities' natural structure and function. It leads to changes in the composition and characteristics of the biotopes. Alterations in the hydrological regime and the geomorphology of water bodies are also reported in the literature. One of the severe consequences of this effect is on the fish habitat, which decreases the fish population. In addition, recreational values also decrease; this may result in other ecological, economic, and socioeconomic consequences.

1.3   MARINE POLLUTION

Marine pollution includes those that arise from offshore oil and gas production and marine oil transportation. Pollutants quickly spread over a considerable distance from the source in the open sea, unlike in the case of soil, where they are fixed to a specific location. The most undesirable aspect of marine pollution is that it is too late to take any corrective measures when it happens. Marine pollutants can be grouped in the increasing order of hazard, as follows:

TABLE 1.1
World's Energy Resources

Source of energy	In 1989	Optimal in 2030
Oil	33	14
Coal	24	8
Gas	18	18
Renewable sources	20	60
Nuclear power	50	0

TABLE 1.2

Anthropogenic Impacts on the Hydrosphere by Offshore Oil and Gas Production

Activity	Sanitary-Hygienic			Ecological			Fisheries		
	L	R	G	L	R	G	L	R	G
Liquid & solid waste discharge	X	X	X	Weak	Weak	X	Weak	Weak	X
Subsea pipelines	X	X	X	considerable	uncertain	X	considerable	High	X
Offshore structure abandoned	X	X	X	Weak	Weak	X	considerable	considerable	X
Accidents causing chemical pollution	considerable	Weak	X	Very high	Weak	X	Very high	Weak	X

G—global; L—local; R—regional.

TABLE 1.3

Anthropogenic Impacts on Hydrosphere on Land Oil and Gas Production

Activity	Sanitary-Hygienic			Ecological			Fisheries		
	L	R	G	L	R	G	L	R	G
Oil pollution	Considerable	X	X	Considerable	Weak	X	Considerable	Weak	X
Subsea pipelines causing chemical pollution	Very high	X	X	Very high	X	X	Very high	Weak	X

G—global; L—local; R—regional.

- *Group 1* refers to those substances causing mechanical impacts that damage respiratory organs, digestive systems, etc. For example, suspensions, films, solid wastes, etc.
- *Group 2* refers to those substances that provoke eutrophic effects and result in the rapid growth of phytoplankton. It causes disturbance of eco-structure and affects various functions of water ecosystems. For example, mineral compounds, organic substances, etc.
- *Group 3* includes those substances with saprogenic properties (sewage with a high content of quickly decomposing organic matter), which causes oxygen deficiency.

- *Group 4* includes substances causing toxic effects, which cause irreversible damage to physiological processes and functions of reproduction. For example, heavy metals, chlorinated hydrocarbons, etc.
- *Group 5* includes those substances that cause carcinogenic, mutagenic, and teratogenic effects. For example, benzo(a)pyrene and other polyclinic aromatic compounds, biphenyls, etc.

Various components that are responsible for marine pollution and their scale of distribution are given in Table 1.4.

Different factors contribute to the estimate of consequences of marine pollutants; namely, (i) hazardous properties of the contaminants; (ii) volume of their input into the ocean; (iii) scale of distribution; (iv) pattern of their behavior in ecosystems; and (v) stability of their composition (Engelhard et al., 1994; Frank and Morgan, 1979). Worldwide contaminants of marine pollutants are given in Table 1.5. The

TABLE 1.4

Scale of Marine Pollution Components

Type of Impact	Scale of Distribution	Sanitary	Eco-Fisheries	Sources
Oil slicks, tar balls	Local	Considerable	Considerable	Oil production and transportation
Suspended solids	Local, regional	Considerable	Considerable	Bottom dredging,offshore structure emplacement, drilling
Oil hydrocarbons: crude oil and oil products	Local, regional, global	Considerable	Considerable	Oil production, storage, marine transportation
Hydrocarbons of methane series	Local, regional	Weak	Considerable	Natural gas production

TABLE 1.5

Level of Contaminants in µg/l in Surface Waters

Ecological Zone	Oil Hydro carbons	Chlorinated Hydrocarbons	Mercury	Lead	Cadmium
South zone	$<10^{-1}$ to 1	$<10^{-4}$ to 10^{-3}	10^{-4} to 10^{-2}	10^{-3} to 10^{-2}	10^{-4} to 10^{-2}
Ocean pelagic area southern part	$<10^{-1}$ to 1	$<10^{-3}$ to 10^{-2}	10^{-4} to 10^{-2}	10^{-3} to 10^{-2}	10^{-4} to 10^{-1}
Enclosed sea open waters	<1 to 10^{-2}	$<10^{-3}$ to 10^{-1}	$<10^{-3}$ to 10^{-2}	10^{-3} to 10^{-1}	10^{-3} to 10^{-1}
Coastal zones	10 to 10^{2}	10^{-3} to 1	10^{-3} to 10^{-1}	10^{-2} to 1	$<10^{-2}$ to 10^{-1}

anthropogenic impact on the ocean environment causes a cumulative effect on oil and gas production facilities. It can be seen from the table that their effects are primarily dose-response in a linear relationship, which is the leading factor for the anthropogenic impact on marine ecosystems (GESAMP, 1991). Offshore activities contribute about 2–5% of the overall pollution in the ocean environment. Also, anthropogenic impact increases the concentration on marine coastal areas and shelf zones.

1.4 IMPACT CAUSED BY OIL AND GAS INDUSTRIES

It is necessary to revisit various oil and gas exploration and production stages to understand the impact caused by the exploration and production of hydrocarbons. The oil and gas production operation has the following stages:

1. *Exploration* is essential for identifying the rig placement, exploratory drilling, plugging the well, killing production wells, etc.
2. *Development and production* are two main stages: platform commissioning, pipeline laying, production drilling, pipeline maintenance, etc.
3. *Decommissioning* is the final stage of the oil and gas production. It includes the removal of the platform, well plugging, etc., when the well is drained.

Environmental impact on each stage of oil and gas development is given in Table 1.6, while oil discharges in the North Sea are shown in Table 1.7.

A typical drilling fluid handling system is shown in Figure 1.1. Various complexities involved in the drilling operation result in a high probability of pollution to the marine environment. It is important to note that, while every care is taken in designing an efficient drilling system, consequences arise in various stages that cause severe impacts on the marine environment. These are accidental and can neither be

TABLE 1.6
Impact on Each Stage of Oil and Gas Production

Stage	Activity	Nature of Impact
Geological and geographical survey	Seismic surveys	Interference with fish and other living organisms
	Test drilling	Sediment re-suspension, increase in turbidity
Exploration	Rig placement	Pollutants discharge
	Exploratory drilling	Interference with fisheries
	Platform placement	Physical disturbances
Development and production	Pipeline laying	Accidental discharges
	Drilling of production well	Physical disturbances
	Support vessel traffic	Disrupting marine birds
		Operational emissions
De-commissioning	Platform removal	Residual remains
	Plugging of wells	Impact when explosives are used
		Operational discharges

TABLE 1.7

Oil Discharge in the North Sea

Description	Oil Discharge in tons/year					
	1984	1985	1986	1988	1989	1990
Drilling cuttings	23 000	26 000	20 000	22 000	16 000	14 000
Diesel-based drilling	2100	Not available				
Drilling discharge	2000	4000	4000	6000	4000	6000
Accident spills	1000	1000	5000	4000	1000	2000

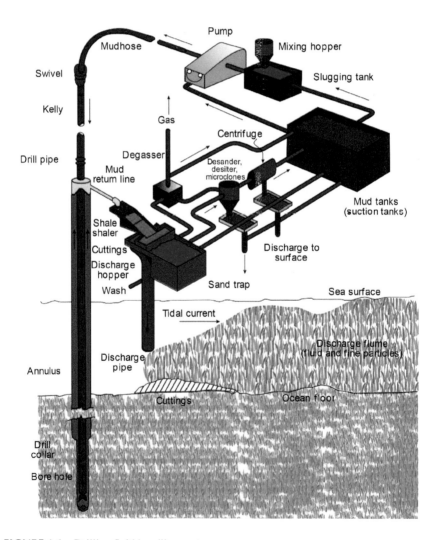

FIGURE 1.1   Drilling fluid handling system.

predicted nor avoided. So, the whole effort is toward reducing the consequences of such events, which are highly unexpected during drilling operations.

1.5 DRILLING OPERATIONS AND CONSEQUENCES

Periodic discharge of drilling mud from a single well is about 15–30 lb, while the mud cuttings contain a dry mass of about 200–1000 tons from a single well; this is still more in the case of multiple wells. Wastewater discharge is about 1500 tons per day from a single production platform. Volumes of discharge in the ocean in different parts of the world are given in Table 1.8.

It is necessary to know the main constituents of oil-based drilling fluid to understand the consequences caused by drilling discharge in the open sea. They are as follows:

Barite: 409 tons (61%)

It is one of the main constituents of the drilling mud and can change surface sediments' texture and erosion properties near offshore drilling sites.

Base oil: 210 tons (31%)

Base oil in use today is formulated with diesel, mineral oil, or low-toxicity linear olefins and paraffin. The olefins and paraffin are often referred to as synthetics, although some are derived from distillation of crude oil, and some are chemically synthesized from smaller molecules.

Calcium chloride: 22 tons (3.35%)

Calcium chloride is a common soluble salt used in drilling and toward the completion stages. It will be powder, pellet, or granular. Calcium chloride is highly hygroscopic, and hence, appropriate protection is essential.

Emulsifier: 15 tons (2.2%)

Emulsifiers may cause emulsion blockage. It will result in increased viscosity and thus impair the mobility of crude oil.

Other constituents **include filtrate agent: 12 tons (1.8%); lime: 2 tons (0.25%); and viscosities: 2 tons (0.4%).**

Each component of the drilling fluid has at least one severe technological effect. Drilling discharge contains heavy metals that severely impact the marine environment.

TABLE 1.8

Volume of Discharge in the Ocean in Different Parts of the World

Country	Volume of Discharge (m³/day)
U.S., GoM	550 000
Offshore California	14 650
Cook Inlet, Alaska	22 065
North Sea	512 000
Australia	100 000

1.6 POLLUTION DUE TO PRODUCED WATERS

Produced water during drilling operations contains dissolved salts and organic compounds. Oil hydrocarbons, trace metals, and suspensions are also present, making the composition of produced water very complex. Produced water generally contains benzene, toluene, and xylenes (10–30 mg/kg). *Biocides, which are used to control biological activities with limited efficiency, are also present*. Organic molecules and heavy metals in produced water are essential sources of marine pollution. Chromatographic analysis of the discharged water at the Gulf of Mexico (GoM) showed higher and relatively stable levels of phenol and its alkylated homologs in the drilling discharges. Even radioactive elements like radium-226 and radium-228 are seen in produced waters. Though at a low level, radioactive elements remain the focus of marine pollution. During contact with seawater, these radionuclides interact with sulfates and residue to form a radioactive scale. They increase radioactive risk in the local and regional areas of produced waters (OGP, 2010; OISD, 2002, 2005, 2011, 2013; Ostgard and Jensen, 1983). It affects the marine life significantly (Ale, 2002; Aven and Vinnem, 2007; Chandrasekaran, 2010a, 2010b, 2011a, 2011b, 2011c, 2011d, 2011e, 2014b, 2014c, 2015a; Papazoglou et al., 2003; Pasman et al., 2009).

1.7 DRILLING ACCIDENTS

Drilling accidents occur due to unexpected blowouts of liquid and hydrocarbons from the well, resulting in significant oil spills (Bonvicini et al., 1998; Bottelberghs, 2000; Brazier and Greenwood, 1998). One of the world's largest oil spills occurred in 1979 near the shore of Mexico after the blowout of drilling rig Ixtoc-1. It resulted in an oil spill for about ten months; the quantity of oil spills ranged from 2500 to 6000 tons. Drilling accidents can be classified into two types; namely, (i) cases leading to a catastrophic situation involving intense and prolonged hydrocarbon gushing and (ii) cases leading to routine hydrocarbon spills and blowouts that occur during normal drilling operations. Although the latter is not reported frequently, it is responsible for causing severe pollution to the marine environment.

1.7.1 UNDERWATER STORAGE RESERVOIRS

Underwater reservoirs are used to store liquid hydrocarbons in large volumes. They are used when tankers are deployed for oil transportation instead of pipelines (Guo et al., 2005; Guo, 2007; Gusto, 2010). Due to the storage space limitation on the topside of drilling and production platforms, underground reservoirs are commonly used. Unfortunately, risks arising from the damage of such reservoirs caused by the collision of vessels and tugboats are also exceptionally high (Chandrasekaran and Kiran, 2014a, 2014b, 2015; Chandrasekaran et al., 2010). Such accidents usually occur during the tanker loading process or due to severe weather conditions. On damage, they become a concentrated source of marine pollution and are toxic with high methanol concentrations (Chandrasekaran, 2013b; Katbas and Aktan, 2002; Khan and Abbasi, 1999; Kiran, 2012; Kiran, 2014; Kletz, 2003). Though the spread of such pollution is comparatively simple, its increased concentration and severe consequences on aquatic life in the closer vicinity are a subject of concern.

1.7.2 PIPELINES

Pipelines transport the explored crude oil to the shore for further processing. Owing to the increased complexities that arise from the failure of pipelines, oil industries are keen on investing in modern regasification plants as a part of the production facility itself. The extensive length of pipelines and inaccessibility for periodic inspection are the main reasons for the potential source of failure. Other factors that cause environmental risk during offshore developments due to pipelines are material defects, corrosion, tectonic movements, and encountering ship anchors and bottom vessels (Bhattacharyya et al., 2010a, 2010b; Vanem et al., 2008; Venkata Kiran, 2011). The pipeline can cause a short- to long-term leakage, which can remain a potential threat over some time. The intensity and scale of toxic impacts released by the failure of pipelines vary, depending upon the combination of these factors.

1.7.3 IMPACT ON MARINE POLLUTION

The large and multi-scale activity of the offshore oil and gas industry imposes a severe impact on the marine environment. It is a significant cause of concern among the environmentalists. The effects of marine pollution are chemical, physical, and biological. Physical hazards can arise during marine surveys (IEC 61882; IS code, 2006). Seismic signals generated during marine surveys are hazardous to the marine fauna, while the explosive activities of abandoned platforms result in the mass migration of commercial fish. Chemical pollution is one of marine pollution's primary and most significant impacts. Significant offshore accidents that resulted in oil spills in the past have led to severe ecological consequences. Moreover, the fate of unused oil platforms and underwater pipelines causes a severe threat to marine ecology, which is one of the passive consequences of offshore drilling activities.

1.8 OIL HYDROCARBONS: COMPOSITION AND CONSEQUENCES

Abundant evidence from the published reports demonstrates the global distribution of oil contamination, primarily from offshore platforms. Concerns about the scale and consequences of oil pollution have increased over the last few decades. Oil input is undoubtedly seen as one of the severe threats to the marine environment (Johnson and Cornwell, 2007; Kakuno, 1983). Crude oil that contains hydrocarbons with a few hybrid compositions, such as paraffin-naphthenic, naphthene-aromatic, etc., is a potential source of marine pollution. Their behavior and biological impact on ecosystems are generally governed by physical and physiochemical properties such as specific gravity, volatility, and water solubility. A vast difference in the properties of oil components leads to the physical fractionalizing of crude oil in the ocean environment. It makes the oil present in different physical states: (i) as surface films, (ii) in the dissolved forms, (iii) as emulsion (oil-in-water, water-in-oil), (iv) in the suspended forms, and (v) as oil aggregates that float on the surface. The suspended particles absorb their fractions; solid and viscous components get deposited at the sea bottom; and other compounds accumulate in the water organisms.

1.8.1 DETECTION OF OIL CONTENT

One of the main problems of detecting the presence of oil content in marine pollution is the existence of hydrocarbons similar to those produced by marine living organisms. Detection becomes even more complicated when the oil presence is low but has high background concentrations. The complex process of oil transformation starts developing when oil comes in contact with seawater. However, the progression, duration, and results of these transformations depend on the properties and composition of the crude oil. Oil spread on the free surface occurs under the influence of gravitational forces.

1.9 OIL SPILL

Oil spills undergo various stages, each significantly polluting the marine environment. Various stages include the following: (i) physical transport, (ii) microbial degradation, (iii) aggregation, and (iv) self-purification. It can disperse over a circumference along the free surface within a few minutes of the oil spill. It forms a thin slick of about 10 mm thick; thinner as it spreads further. Spread of oil spills can even extend to a few square kilometers. A considerable part of the spillage transforms into the gaseous phase during the first few days after the spill. While the slick gradually loses water-soluble hydrocarbons, the remaining fraction, being viscous, reduces slick spreading. Most of the oil components, like aliphatic and aromatic hydrocarbons, are water-soluble to a certain degree due to their lower molecular weight. Hydrodynamic and physiochemical conditions influence oil dissolution rate in surface waters. The chemical transformations of oil on the water surface occur as early as the day of the oil spill, resulting in the reaction's oxidative nature. It involves petrochemical responses under the influence of ultra violet waves of the solar spectrum. Some traces of vanadium and compounds of sulfur catalyze the oxidation process. The final oxidation products, such as hydroperoxides, phenols, carboxylic acids, ketones, aldehydes, and others, have increased water solubility and toxicity.

1.9.1 ENVIRONMENTAL IMPACT

Available data on seawater levels of oil hydrocarbons vary in different regions. Factors influencing them arise from the complexities of their bio-geo-chemical behavior. Reports show that the tendencies of oil levels tend to increase from the ocean pelagic region to the enclosed sea, coastal waters, and estuaries. Marine pollution studies also identified the maximum contamination of the euphotic layer, patchy distribution of contaminants, localization in the upper microlayer, deposition in bottom sediments, increased levels in contact zones, and overlapping fields of maximum pollution through their recent reports

1.9.2 OIL: A MULTICOMPONENT TOXICANT

The eco-toxicological characteristics of oil are highly complex and cause variability in its composition. Oil is an essential toxicant due to its integrated nature. It affects every vital function, such as process, mechanism, and system of living organisms. Oil hydrocarbons with complex molecules are more toxic than simpler molecules and a straight chain of carbon atoms. Increasing the molecular weight of the components increases their toxicity. Biomarker methodology is an essential element of marine monitoring, which provides data for assessing the cumulative biological effects of the chronic oil contamination of seawater.

1.9.3   OIL SPILL

Oil hydrocarbons are continuously released in the marine environment due to natural oil seepage from the sea floor. The global distribution of oil hydrocarbons in the World Ocean is characterized by increasing concentrations from pelagic areas to coastal waters. From the chemical point of view, oil is a complex mixture of many organic substances, which are dominated by hydrocarbons. When they come in contact with the marine environment, they are easily separated into fractions. These separated fractions form surface slicks, dissolved and suspended substances, emulsions, and solid and viscous components. Migration of oil, from a biological perspective, is a complex and interconnected process. They include physical transport, dissolving and emulsification, oxidation and decomposition, sedimentation, and microbial degradation.

1.10   CHEMICALS AND WASTES FROM OFFSHORE OIL INDUSTRY

1.10.1   DRILLING DISCHARGES

Drilling mud is hazardous, due to its persistence in marine environments. After six months of discharge of oil-based drilling waste, it was found that they biodegraded only by 5% (Ostgaard and Jensen, 1983). Drilling waste, such as fatty acids, lose their organic fraction due to microbial and physiochemical decomposition. Water-based drilling mud is generally disposed of overboard, which adds more intensity to marine pollution. Drill cuttings, which are pieces of rock crushed by the drill bit, are brought to the surface; they generally do not pose any particular threat but increase the turbidity and smothering effect of benthic organisms. Oil-based mud, in particular, contains a wide array of hazardous organic and inorganic traces. Discharge of drilling cuttings in large volumes imposes eco-toxicological disturbances in offshore production areas. Oil and the oil products in the drilling cuttings are the main toxic agents. The permissible limit of drilling cuttings discharge cannot exceed 100 g/kg; in reality, this concentration is exceeded by about 100 times. Drilling waste discharged into the marine environment disperses in the solid phase. It contains clay minerals, barite, and crushed rock. Large and heavy particles are rapidly sedimented, while the small fractions gradually spread over more considerable distances.

Produced water is one of the forms of discharge that is evacuated from offshore platforms, the volume of which is significantly high. They include solutions of mineral salts, organic acids, heavy metals, and suspended particles. The produced water causes more complications, due to its mixed chemical composition if combined with the injection water used for oil recovery.

1.11   CONTROL OF OIL SPILL

Most drilling fluids' LC_{50} values (96-hour exposure) vary from 10 to 15 g/kg, showing lethal substances of high toxicity. Drilling fluid contains three main groups of toxicity. Group 1 refers to low-toxic substances such as bentonite, barite, and lignosulfonates. Group 2 refers to high-toxic compounds such as biocides, corrosion inhibitors, and descalers; they are seen in small proportions. Group 3 refers to the medium-toxic compounds such as lubricating oil, emulsifiers, thinners, and solvents, which are seen in large percentages. Oil spills can be controlled by many mechanical, chemical, and biological methods; mechanical methods are generally preferred due to higher efficiency. One of the most common mechanical methods of preventing oil spills is deploying floating booms. Oil slick spreading can be controlled using the booms, which are then collected from oil collectors. Special ships with floating separating units are used for this purpose. Usually, mechanical means are supplemented by chemical spill-control methods as well.

1.12   ENVIRONMENTAL MANAGEMENT ISSUES

Environmental issues arise from the oil and gas development activities, which are the current focus of the scientific community; it also draws public attention worldwide. Environmental management policies are framed by the local and global regulatory authorities, which consider the factors of current and future interest. These factors include (i) possibilities of alternative sources of energy, (ii) natural conditions, (iii) ecological factors, and (iv) techno-economic factors.

1.12.1   ENVIRONMENTAL PROTECTION: PRINCIPLES APPLIED TO OIL AND GAS ACTIVITIES

Various factors contribute to the implementation of policies and regulations that are regulatory to control environmental pollution that arises from oil and gas exploration. They are listed as follows:

- Acknowledgment of socioeconomic stipulation: Many countries are framing policies cooperating with oil producers, fishermen, and environmentalists to achieve mutual understanding across their respective domains.
- Expediency of developing offshore natural resources.
- Using an ecocentric approach in contrast with the anthropocentric approach: This alternative approach ensures the stability of natural ecosystems. It supports conditions for self-renewal of biological resources.
- Environmental protection policies are governed by regional aspects accounting for specific features of different marine basins in terms of diverse climate, social, economic, and other characteristics.

Guidelines framed by the Joint Group of Experts on Scientific Aspects of Marine Pollution are generally followed (GESAMP, 1991). These guidelines indicate three main blocks: planning, assessment, and regulation. They include the regulatory measures for discharging drilling waste into the sea. The most important guideline, implemented with strict compliance, is that discharge into the sea requires proper authorization. The concentration of oil and oil products determined using standard tests should be, at most, the established standards. LC_{50} values for discharge samples established using a 96-hour Mysid toxicity test should not exceed 30 g/kg (Cano and Dorn, 1996).

1.12.2   Environmental Management: Standards and Requirements

The standards that govern the implementation of environmental management policies include the following:

- The mercury and cadmium content in the drilling fluid's barite base is restricted.
- Drilling waste is not discharged in waters within 3 miles of the shore.
- No discharge of diesel oil is allowed.
- No discharge of free oil, based on the static sheen test, is allowed. Tests should conform to LC_{50} values based on the mysid toxicity test.
- Average oil concentration should not be more than seven mg/l for a monthly oil content and 13 mg/l for an average daily oil concentration.
- Discharge must be measured within 4 miles of the shore to ascertain toxicity compliance.

1.13   ECOLOGICAL MONITORING

Ecological monitoring is a system that collects information about the changes in natural parameters due to oil pollution in the open sea. Since the contents of oil pollution cannot be measured directly due to its complex composition, its effects are measured in terms of its consequences on marine organisms; this is an indirect monitoring method. Ecological monitoring is one of the primary methods to control and manage activities related to marine pollution. Biological monitoring is based on measuring molecular and cellular effects under low levels of impact that are incapable of chemical analysis. Ecological monitoring in offshore oil production is done at the local level. The results of the monitoring should be strictly in compliance with the established standards.

1.13.1   Ecological Monitoring Stages

Ecological monitoring is done in different stages. In the first stage, potential hazards from the impact sources are identified (Kyriakdis, 2003; Lees, 1996; Leffler et al., 2011; Leonelli et al., 1999). In the second stage, regular observations of marine biota are made to assess the responses in biological organisms qualitatively. The cause-effect relationship between the biological effects and impact factors is then studied. Subsequently, the total impact on the marine environment and biota, including the effects on commercial species and biological resources, was assessed. At the final stage, corrective measures are recommended for checking marine pollution and suitable preventive measures, if any.

1.14 ATMOSPHERIC POLLUTION

1.14.1 RELEASE AND DISPERSION MODELS

The release model identifies the type of release of material (Webber et al., 1992; Wiltox, 2001). It assesses the release rate of toxic material into the atmosphere. It also includes estimation of the downwind concentration of the released material. *The dispersion model* describes how vapors are transported downwind of a release. Three different kinds of vapor cloud behavior and the corresponding release time models are considered, which are given in Table 1.9.

TABLE 1.9
Vapor Cloud Behavior

Vapor cloud behavior	Release-time mode
Neutrally-buoyant gas	Puff model
Positively-buoyant gas	Plume model
Dense-buoyant gas	Time-varying model (continuous)

1.14.2 PLUME AND PUFF MODELS

Continuous release source   *Place*

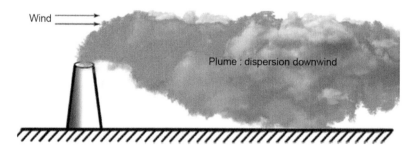

FIGURE 1.2   Continuous release (plume).

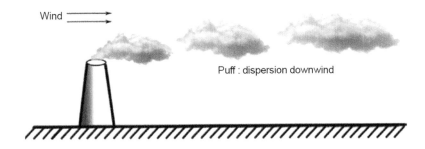

Instantaneous release source   Place and time

FIGURE 1.3   Instantaneous release (puff).

Parameters that affect dispersion are wind speed, terrain effects, atmospheric stability, height of release above ground, and the initial momentum of the released material.

1.14.2.1   Wind Speed

Dispersed or emitted gas initially dilutes with air passage; emitted gas is carried forward faster (Chuhan-Jie et al., 2013; Crawley et al., 2000). In this process, it also gets diluted due to adding a large quantity of air. Wind speed and direction can be obtained from the wind rose diagram for a particular geographic location. The wind rose diagrams are plotted for specific regions, showing wind speed, direction, and relative frequency. It comprises 16 angular wedges, each representing an arc of 22.5° segments. It contains eight colored segments, each representing a specific wind speed when blowing from a particular direction. The overall radius of each wedge represents the percentage of time wind came from that direction during the period of interest. Wind speed and the direction of wind influence the release significantly. The near-neutral and stable air condition for the wind profile is given by:

$$u_z = u_{10}\left(\frac{Z}{10}\right)^p \tag{1.1}$$

Where p is the power coefficient ($p = 0.4$, 0.28, and 0.16 for urban, suburban, and rural areas, respectively), U_{10} is the wind speed at 10 m elevation, and Z is the elevation (in m).

1.14.2.2   Terrain Effects

Ground conditions and terrain effects also influence the mechanical mixing at the surface, while height affects the wind profile. Physical interferences caused by trees and buildings increase the blending, but open sources like lakes and open ground decrease the mixing. The wind profile in different areas is given in Figure 1.4.

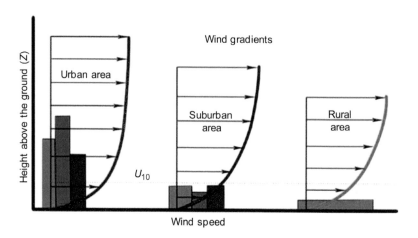

FIGURE 1.4   Wind profile for different areas.

1.14.2.3   Atmospheric Stability

The atmospheric vertical temperature gradient defines it. During the daytime, air temperature decreases rapidly with the increase in height; this will encourage the vertical lift motion. The lapse rate is given by:

$$T = -\left(\frac{dT}{dZ}\right) \cong 1^{\circ}C/100m \tag{1.2}$$

Where dT is the temperature differential, and dZ is the variation in height.

Atmospheric stability is classified into unstable, neutral, and stable. Under unstable atmospheric conditions, the sun heats the ground faster than the heat that can be removed so that air temperature near the ground is higher than that at higher elevations. Under neutral atmospheric conditions, the air above the ground warms, increasing the wind speed. It reduces the effect of solar input. Under stable atmospheric conditions, the sun cannot heat the ground as fast as the ground cools. As a result, the temperature at the ground will be comparatively lower. Figure 1.5 shows the air temperature as a function of altitude for day and night conditions. Plots consider the adiabatic temperature gradient for humid air to be 0.5°C/100 m.

The atmospheric stability classes are also classified according to the Pasquill Stability of the atmosphere. It is classified into six categories: A, B, C, D, E, and F. A is a precarious condition with deficient wind speed. B is a moderately unstable condition. Atmospheric stability class C refers to a slightly stable condition with increased wind velocity. In contrast, class D refers to a neutrally stable condition generally used for overcast conditions. Class E is a somewhat stable condition generally used for night conditions, while class F refers to a moderately stable atmospheric condition. Table 1.10 shows the Pasquill Stability classes for day and night conditions.

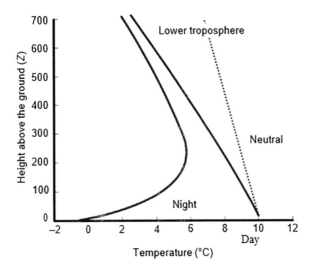

FIGURE 1.5   Air temperature as a function of altitude.

TABLE 1.10
Pasquill Stability Classes for Day and Night Conditions

Surface Wind Speed (m/s)	Day, Incoming Solar Radiation			Night, Cloud Cover Thickly Overcast		Amiime Heavy Overcast
	Strong	Moderate	Slight	>1/2 low clouds	<3/8 clouds	
<2	A	A-B	B	F	F	D
2-3	A-B	B	C	E	F	D
3-5	B	B-C	D	D	E	D
5~6	C	C-D	D	D	D	D
>6	C	D	D	D	D	D

1.14.2.4   Height of Release above Ground and Momentum of Material

The ground level concentration of a dispersed plume decreases with the increase of the source of the release height. Figure 1.6 shows the release above the ground. The momentum of the released material depends on the effective release height and initial buoyancy. For example, the momentum of a high-velocity jet will carry the released material with a velocity higher than that at the point of release. Gas will be initially negatively buoyant and will slump toward the ground. If gas has a lower density than air, it will initially be positively buoyant and will be lifted upward.

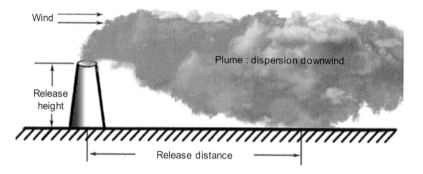

FIGURE 1.6 Release of material above ground.

1.15 DISPERSION MODELS FOR NEUTRALLY AND POSITIVELY BUOYANT GAS

Neutrally and positively buoyant gas dispersion models help estimate the average concentrations and predict the time profile of flammable, toxic gases along the downwind direction of the release. Like those liquid release models, plume and puff models are commonly used to model the vapor cloud dispersion. The plume model describes the continuous emission of materials from a steady height, H, above the ground level, shown in Figure 1.7. The wind-blowing direction is taken along the X-axis.

1.15.1 PLUME DISPERSION MODELS

The average released material or gas concentration is given by:

$$C(x,y,z) = \frac{Q}{2\Pi\sigma_x\sigma_y U}\exp\left[-\frac{1}{2}\left(\frac{y}{\sigma_y}\right)^2\right]x$$
$$\left\{\exp\left[-\frac{1}{2}\left(\frac{z-H}{\sigma_z}\right)^2\right]+\exp\left[-\frac{1}{2}\left(\frac{z+H}{\sigma_z}\right)^2\right]\right\} \tag{1.3}$$

where $C(x, y, z)$ is the average concentration of release material (kg/m3), H is the height of the releasing source from the ground (m), (x, y, z) are distances from the source in down-wind, cross-wind, and vertical direction, respectively (m), Q is the release strength of the material (kg/s), U is the wind velocity (m/s), and $(\sigma y, \sigma z)$ are dispersion coefficients in y and z directions, respectively.

Let us consider some of the cases of the plume dispersion model, as discussed later:

Case 1: Ground level centerline concentration ($y = z = 0$), then the concentration is given by:

$$C(x,0,0) = \frac{Q}{2\Pi\sigma_z\sigma_y U}\exp\left[-\frac{1}{2}\left(\frac{H}{\sigma_z}\right)^2\right] \tag{1.4}$$

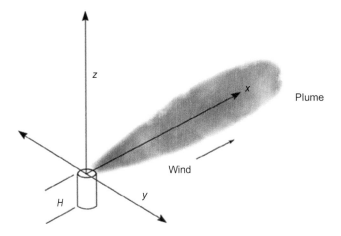

FIGURE 1.7   Plume dispersion model.

Case 2: Ground, centerline, release height, $H = 0$, then the concentration is given by:

$$C(x,0,0) = \frac{Q}{2\Pi\sigma_z\sigma_y U} \tag{1.5}$$

In both cases, x is implicit with the dispersion coefficients.

1.15.2   MAXIMUM PLUME CONCENTRATION

Maximum plume concentration always occurs at the release point (Brode, 1959). For releases above ground, maximum concentration occurs along the centerline (*X-axis*) of the downwind direction. The distance at which maximum ground level concentration occurs is given by:

$$\sigma_z = \frac{H}{\sqrt{2}} \tag{1.6}$$

The maximum concentration is given by:

$$C_{max} = \frac{2Q}{e\Pi U H^2}\left(\frac{\sigma_z}{\sigma_y}\right) \tag{1.7}$$

1.16   PUFF DISPERSION MODEL

The puff dispersion model describes the instantaneous release of the material (e.g., a sudden chemical release from a ruptured vessel). The consequence of such a release will be the forming of a large vapor cloud from the dispersed (rupture) point. In this case, the classical puff model is also used to describe a plume. The average concentration is estimated for the puff release using the following relationship:

$$C(x,y,z) = \frac{Q_{instantaneous}}{(2\Pi)^{3/2}\sigma_x\sigma_y\sigma_z}\exp\left[-\frac{1}{2}\left(\frac{x-ut}{\sigma_x}\right)^2\right]x$$
$$\exp\left[-\frac{1}{2}\left(\frac{y}{\sigma_y}\right)^2\right] + \left\{\exp\left[-\frac{1}{2}\left(\frac{z+H}{\sigma_z}\right)^2\right] + \exp\left[-\frac{1}{2}\left(\frac{z+H}{\sigma_z}\right)^2\right]\right\}$$

(1.8)

Let us consider some exceptional cases of puff modeling.

Case 1: The total integrated dose at ground level (i.e., $z = 0$) is given by:

$$\text{Dose}(x,y,0) = \frac{Q_{instantaneous}}{\Pi\sigma_y\sigma_z u}\exp\left[-\frac{1}{2}\left(\frac{y}{\sigma_y}\right)^2 - \frac{1}{2}\left(\frac{H}{\sigma_z}\right)^2\right]$$

(1.9)

Case 2: The concentration on the ground below the puff center is given by:

$$C(x,0,0,t) = \frac{Q_{instantaneous}}{\sqrt{2}\Pi^{3/2}\sigma_z\sigma_y\sigma_y}\exp\left[-\frac{1}{2}\left(\frac{H}{\sigma_z}\right)^2\right]$$

(1.10)

Case 3: Puff center on the ground (i.e., $H = 0$) is given by:

$$C(x,0,0,t) = \frac{Q_{instantaneous}}{\sqrt{2}\Pi^{3/2}\sigma_z\sigma_y\sigma_y}$$

(1.11)

1.16.1 MAXIMUM PUFF CONCENTRATION

The maximum puff center is located at the release height, and the center of puff is located at $x (= ut$, where u is the wind velocity). On the ground, maximum concentration always occurs directly below the puff center.

1.17 ISOPLETHS

Isopleths measure the cloud boundary at a fixed concentration. It represents the lines of constant concentration, as shown in Figure 1.8. Different steps to determine isopleths are as follows:

Step 1: Determine concentrations along the centerline at fixed points along the downwind direction.

Step 2: Find the off center distances to isopleths (y) at each point using Equation (1.12).

$$y = \sigma_y\sqrt{2\ln\left[\frac{C(x,0,0,t)}{C(x,y,0,t)}\right]}$$

(1.12)

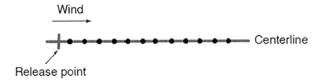

FIGURE 1.8 Isopleths in wave propagation.

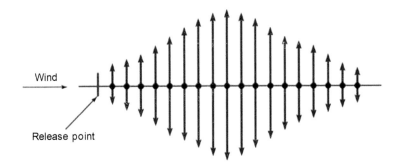

FIGURE 1.9 Isopleths offset.

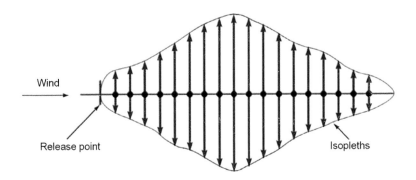

FIGURE 1.10 Isopleths in transverse spread.

where $C(x, 0, 0, t)$ is the down-wind ground centerline concentration and $C(x, y, 0, t)$ is the isopleth concentration at (x, y).

  Step 3: Plot isopleths offset for both directions at each point, as shown in Figure 1.9.

  Step 4: Connect the points shown in Figure 1.10 to get isopleths.

1.18 DISPERSION COEFFICIENTS

The dispersion coefficients are essential to model the release scenarios using plume or puff models (Wiltox, 2001). They depend upon the stability class and down-wind distances. To calculate the dispersion coefficients, initially identify the Pasquill Stability class using meteorological data such as wind speed, heat radiation, cloud cover, etc. (Chamberlain, 1987; Chandrasekaran, 2013b, 2015c; Pontiggia et al., 2011; Prem et al., 2010; Zhang and Dong, 2013; Ramamurthy, 2011; Rodante, 2004; Sanderson, 1988). Classification of the area as rural, urban, flat, or hilly is also required. Using Figures 1.11, 1.12, and 1.13, one can estimate the dispersion coefficients for the relevant cases as applicable.

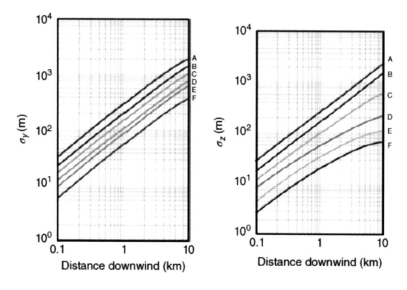

FIGURE 1.11   Dispersion coefficients for plume model (rural release).

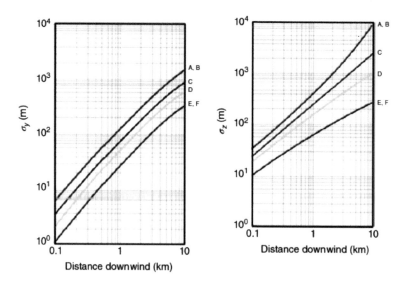

FIGURE 1.12   Dispersion coefficients for plume model (urban release).

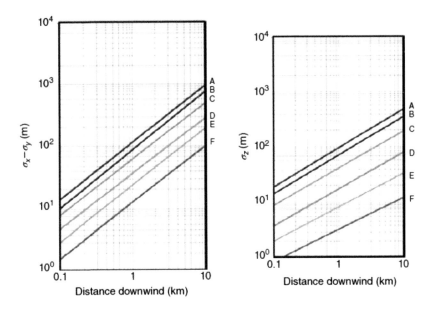

FIGURE 1.13 Dispersion coefficients for puff model.

1.18.1 ESTIMATES FROM EQUATIONS

Dispersion coefficients for the plume model can also be estimated mathematically. Let X be the downwind distance (in m) measured from the source of release. Dispersion coefficients can be calculated in Tables 1.11 and 1.12 for plume and puff models, respectively.

1.19 DENSE GAS DISPERSION

Gases having a density higher than air are termed dense gases. Dense gases released from the source initially slump toward the ground, moving upward and progressing along the downwind directions. Mixing mechanisms with air differs entirely from the neutrally buoyant releases. Britter-McQuaid dense gas dispersion model is commonly used in such cases.

1.19.1 BITTER-McQUAID MODEL

Step 1: Characterize the initial buoyancy using the following relationship:

$$g_0 = g\left|\frac{\rho_0 - \rho_a}{\rho_a}\right| \qquad (1.13)$$

Where g is the acceleration due to gravity, ρ_0, and ρ_a is the density of the released material and ambient air, respectively.

TABLE 1.11

Calculation of Dispersion Coefficients (Plume Model)

Area	Stability Class	σ_y (m)	σ_z (m)
Rural conditions	A	$0.22X(1 + 0.0001X)^{-0.5}$	$0.20X$
	B	$0.16X(1 + 0.0001X)^{-0.5}$	$0.12X$
	C	$0.11X(1 + 0.0001X)^{-0.5}$	$0.08X(1 + 0.0002X)^{-0.5}$
	D	$0.08X(1 + 0.0001X)^{-0.5}$	$0.06X(1 + 0.0015X)^{-0.5}$
	E	$0.06X(1 + 0.0001X)^{-0.5}$	$0.03X(1 + 0.0003X)^{-1.0}$
	F	$0.04X(1 + 0.0001X)^{-0.5}$	$0.016X(1 + 0.0003X)^{-1.0}$
Urban conditions	A-B	$0.32X(1 + 0.0004X)^{-0.5}$	$0.24X(1 + 0.0001X)^{-0.5}$
	C	$0.22X(1 + 0.0004X)^{-0.5}$	$0.20X$
	D	$0.16X(1 + 0.0004X)^{-0.5}$	$0.14X(1 + 0.0003X)^{-0.5}$
	E-F	$0.11X(1 + 0.0004X)^{-0.5}$	$0.08X(1 + 0.0001X)^{-0.5}$

TABLE 1.12

Calculation of Dispersion Coefficients (Puff Model)

Area	Stability Class	σ_x or σ_y (m)	σ_z (m)
Rural conditions	A	$0.18X^{0.92}$	$0.60X^{0.75}$
	B	$0.14X^{0.92}$	$0.53X^{0.73}$
	C	$0.10X^{0.92}$	$0.34X^{0.7'}$
	D	$0.06X^{0.92}$	$0.15X^{0.70}$
	E	$0.04X^{0.92}$	$0.10X^{0.65}$
	F	$0.02X^{0.89}$	$0.05X^{0.61}$

Step 2: Decide whether the release is instantaneous or continuous using the following relationship:

$$F = \left(\frac{uR_d}{x}\right) \tag{1.14}$$

Where u is the wind velocity, x is the distance from the release point, and Rd is the release duration. For $F \geq 2.5$, the release is assumed to be continuous; for $F \leq 0.6$, it is considered an instantaneous release. In case,

$0.6 < F < 2.5$ is satisfied, and then one can use both approaches to find the maximum value.

Step 3: Characterize the source dimension.

For continuous release, source dimension (Dc) is given by:

$$D_c = \sqrt{\frac{q_0}{u}} \tag{1.15}$$

Where q0 is the initial plume volume flux and u is the wind speed. For instantaneous release, source dimension (Di) is given by:

$$D_i = V_0^{1/3} \qquad (1.16)$$

where V_0 is the initial volume.

Step 4: Checking criteria.
For continuous release, the value is checked using the following relationship:

$$\left[\frac{g_0 q_0}{u^3 D_c}\right] \geq 0.15 \qquad (1.17)$$

For instantaneous release, the check is done using the following equation:

$$\frac{\sqrt{g_0 V_0}}{u D_i} \geq 0.20 \qquad (1.18)$$

If the criterion is satisfied, the concentration ratio (Cm/C_0) is given in Figure 1.14. The concentration ratio is also given by the relationships shown in Tables 1.13 and 1.14 for gas plume and gas puff models, respectively.

1.20   DISPERSED LIQUID AND GAS: TOXIC EFFECTS

The toxicity of the dispersed liquid or gas in the atmosphere is measured based on the concentration of dispersion and the duration of exposure (Leonelli et al., 1999; Madsen et al., 2006; Marshall, 1969; Marshall and Bea, 1976). Permissible Exposure Limit (PEL) or Threshold Limit Value-Time-Weighted Averages (TLV-TWA) are very conservative estimates of work exposure. There are six alternate methods of toxic effect evaluation (Che Hassan et al., 2009, 2010):

Method 1: Based on Emergency Response Planning (ERPG). The American Industrial Hygiene Association formulates this. In this, three ERPG values are used, namely ERPG-1, ERPG-2, and ERPG-3.

Method 2: Based on the guidelines recommended by the National Institute for Occupational Safety and Health (NIOSH), toxicity is evaluated. NIOSH recommends standards for Immediately Dangerous to Life and Health (IDLH) that explain the level of acceptable toxicity.

Method 3: This is based on the guidelines the National Research Council, Canada (NRC) recommended. NRC recommends Emergency Exposure Guidance Levels (EEGL) for different exposure durations, namely 1-hour EEGL and 24-hour EEGL.

Method 4: This is based on OSHA's Permissible Exposure Limits (PELs). It includes the Occupational Safety and Health Administration and the U.S. Department of Labor.

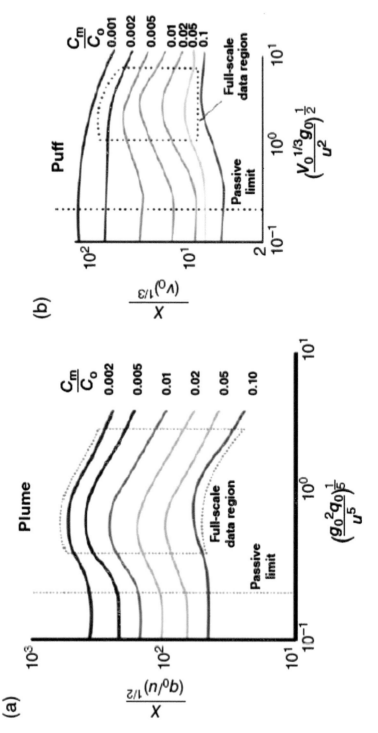

FIGURE 1.14　Concentration ratio: (a) for the plume model; (b) for the puff model.

TABLE 1.13
Dispersion of Dense Gas Plumes

Concentration Ratio (c_m/c_o)	Valid Range for $\alpha = \log(g_o^2 q_e/u^5)^{1/5}$	$\beta = \log[x/(q_o/u)^{1/2}]$
0.1	$\alpha \leq -0.55$	1.75
	$-0.55 < \alpha \leq -0.14$	$0.24\alpha + 1.88$
	$-0.14 < \alpha \leq 1$	$0.50\alpha + 1.78$
0.05	$\alpha \leq -0.68$	1.92
	$-0.68 < \alpha \leq -0.29$	$0.36\alpha + 2.16$
	$-0.29 < \alpha \leq -0.18$	2.06
	$-0.18 < \alpha \leq 1$	$-0.56\alpha + 1.96$
0.02	$\alpha \leq -0.69$	2.08
	$-0.69 < \alpha \leq -0.31$	$-0.45\alpha + 2.39$
	$-0.31 < \alpha \leq -0.16$	2.25
	$-0.16 < \alpha \leq 1$	$-0.54\alpha + 2.16$
0.01	$\alpha \leq -0.7$	2.25
	$-0.7 < \alpha \leq -0.29$	$-0.49\alpha + 2.59$
	$-0.29 < \alpha \leq -0.20$	2.45
	$-0.20 < \alpha \leq 1$	$-0.52\alpha + 2.35$
0.005	$\alpha \leq -0.67$	2.4
	$-0.67 < \alpha \leq -0.28$	$-0.59\alpha + 2.8$
	$-0.28 < \alpha \leq -0.15$	2.63
	$-0.15 < \alpha \leq 1$	$-0.49\alpha + 2.56$
0.002	$\alpha \leq -0.69$	2.6
	$-0.69 < \alpha \leq -0.25$	$-0.39\alpha + 2.87$
	$-0.25 < \alpha \leq -0.13$	277
	$-0.13 < \alpha \leq 1$	$-0.50\alpha + 2.71$

TABLE 1.14
Dispersion of Dense Gas Puffs

Concentration Ratio (c_m/c_o)	Valid Range for $\alpha = \log(g_o V_o^{1/3}/u^2)^{1/2}$	$\beta = \log\left[x/(V_o)^{1/3}\right]$
0.1	$\alpha \leq -0.44$	0.7
	$-0.44 < \alpha \leq 0.43$	$0.26\alpha + 0.81$
	$0.43 < \alpha \leq 1$	0.93
0.05	$\alpha \leq -0.56$	0.85
	$-0.56 < \alpha \leq 0.31$	$0.26\alpha + 1.0$
	$0.31 < \alpha \leq 1$	$-0.12\alpha + 1.12$

(*Continued*)

TABLE 1.14 (Continued)
Dispersion of Dense Gas Puffs

Concentration Ratio (c_m/c_o)	Valid Range for $\alpha = \log\left(g_o V_o^{1/3}/u^2\right)^{1/2}$	$\beta = \log\left[x/\left(V_o\right)^{1/3}\right]$
0.02	$\alpha \leq -0.66$	0.95
	$-0.66 < \alpha \leq 0.32$	$0.36\alpha + 1.19$
	$0.32 < \alpha \leq 1$	$-0.26\alpha + 1.38$
0.01	$\alpha \leq -0.71$	1.15
	$-0.71 < \alpha \leq 0.37$	$0.34\alpha + 1.39$
	$0.37 < \alpha \leq 1$	$-0.38\alpha + 1.66$
0.005	$\alpha \leq -0.52$	1.48
	$-0.52 < \alpha \leq 0.24$	$0.26\alpha + 1.62$
	$0.24 < \alpha \leq 1$	$0.30\alpha + 1.75$
0.002	$\alpha \leq 0.27$	1.83
	$0.27 < \alpha \leq 1$	$-0.32\alpha + 1.92$
0.001	$\alpha \leq -0.10$	2.075
	$-0.10 < \alpha \leq 1$	$-0.27\alpha + 2.05$

Method 5: This is based on the Environmental Protection Agency's (EPA) toxic endpoint. Guidelines recommended by EPA, U.S. EPA/6000/R-7/080 (2007): Sediment Toxicity Identification Evaluation guidelines are followed to estimate the toxicity.

Method 6: This is based on the guidelines the American Conference of Governmental Industrial Hygienists (ACGIH) recommended. ACGIH [1994]. 1994–1995 recommends threshold limit values for chemical substances, physical agents, and biological exposure indices. Cincinnati, OH: American Conference of Governmental Industrial Hygienists.

1.21   HAZARD ASSESSMENT AND ACCIDENT SCENARIOS

The hazard assessment procedure commonly used in oil and gas industries is shown in Figure 1.15. The development of the accident scenario is given in Figure 1.16.

1.21.1   DAMAGE ESTIMATE MODELING: PROBIT MODEL

The probit value, in terms of probability units, is given by:

$$P = \frac{1}{(2\pi)^{1/2}} \int_{-\infty}^{Y-5} exp\left[\frac{-u^2}{2}\right] du \qquad (1.19)$$

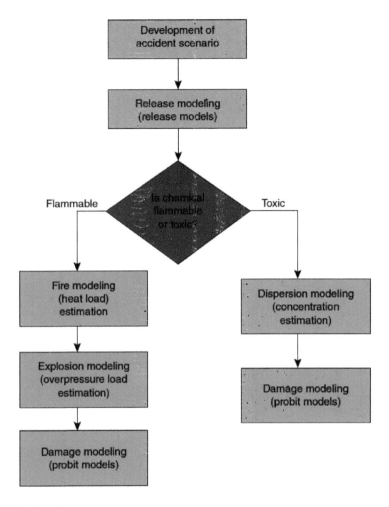

FIGURE 1.15   Hazard assessment.

The probit function transforms the nonlinear dose response into a linear relationship, as shown in Figure 1.17 (also in Eq. (1.20)).

$$\Upsilon = K_1 + K_2 \ln(V) \tag{1.20}$$

K1 and K2 are constants, and V is the dose variable (due to overpressure, radiation, impulse, or dispersion concentration). In a simplified form, probit value (Y) can be transformed to the percentage effect using the following relationship:

$$P = 50 \left[1 + \frac{\Upsilon - 5}{|\Upsilon - 5|} \text{erf}\left(\frac{|\Upsilon - 5|}{\sqrt{2}} \right) \right] \tag{1.21}$$

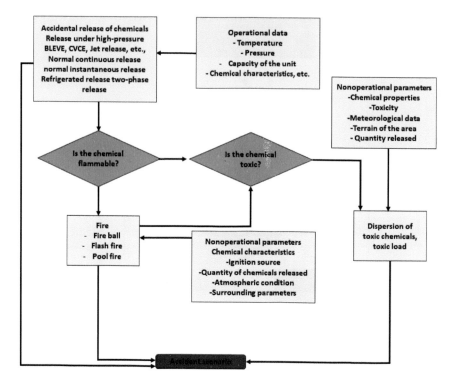

FIGURE 1.16   Development of accident scenario.

where *erf* is the error function.

The probit correlations for various damages are given in Table 1.15.

1.22   FIRE AND EXPLOSION MODELS

The Dow Fire and Explosion Index (FEI) method is used to study fire and explosion releases in oil and gas industries (Papazoglou et al., 2003; Abbasi and Abbasi, 2007; Brode, 1959; Bubbico and Marchini, 2008; Gomez et al., 2008). The flowchart to compute the FEI is given in Figure 1.18.

Step 1: Compute Material factor (M.F.).

The material factor depends on the vapor pressure and flammable (or explosive) characteristics.

Step 2: Compute factor F1.

It depends on general process hazards. For example, hazards arise due to unit operation, such as reaction, material handling, etc.

Step 3: Compute factor F2.

It depends on the particular process hazards. For example, hazards surrounding the unit arise due to special conditions in operation.

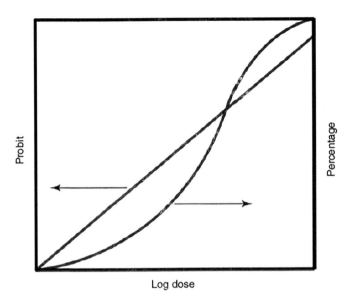

FIGURE 1.17 Probit function.

Step 4: Compute Process Unit Hazard (PUH) using the following relationship:

$$PUH = (1 + F1) * (1 + F2) \qquad (1.22)$$

Step 5: Compute FEI.
The dow FEI is given by:

$$FEI = MF * PUH \qquad (1.23)$$

Based on the computed FEI value, one can rate the degree of hazard, as given in Table 1.16.

1.23 ACCIDENT CASE STUDIES: A BRIEF OUTLINE

1.23.1 SLEIPNER A PLATFORM

Consider an accident reported on the Sleipner A platform in the North Sea. The Sleipner platform is shown in Figure 1.19. It is a condeep-type platform with a concrete gravity base structure, consists of 24 cells, and has a total base area of 16,000 m², operating at a water depth of 82 m. The platform is producing oil and gas successfully in the North Sea. Failure of the platform caused a seismic event of magnitude 3.0 on the Richter scale. The failure resulted in a total economic loss of about $700 million. The conclusions of the investigations mentioned that the failure

TABLE 1.15
Probit Correlations for Various Damages

Type of Damage	Dose Variable	Probit Equation Constants	
		K_1	K_2
Fire			
Burn deaths from fire	$(t*1^{4/3})/10^4$	−14.9	2.56
Explosions			
Deaths from lung hemorrhage	P^0	−77.9	6.91
Eardrum rupture	P^0	−15.6	1.93
Structural damage	P^0	−23,8	2.92
Glass breakage	P^0	−18.1	2.79
Death from overpressure impulse	J	−46.1	4.82
Injuries from overpressure impulse	J	−39.1	4.45
Injures from flying fragments	J	−27.1	4.26
Toxic release and dispersion			
Death due to ammonia dose	$C^{2.0*}T$	−35.9	1.85
Death due to sulfur dioxide dose	$C^{1.0*}T$	−15.67	1
Death due to chlorine dose	$C^{2.0*}T$	−8.29	0.92
Death due to ethylene oxide dose	$C^{1.0*}T$	−6.19	1
Death due to phosgene dose	$C^{1.0*}T$	19.27	3.69
Death due to toluene dose	$C^{2.5*}T$	−6.79	0.41

T is time (s); I is radiation intensity (W/m²); P^0 is peak over pressure (N/m²); J is impulse (N$_s$/m²); C is exposed concentration (ppm); and T is duration of exposure (min).

in a cell wall resulted in a severe crack that propagated. The leakage was so high that the pumps could not control it. The wall failed due to a severe error in the finite element analysis and insufficient anchorage of the reinforcement in a critical zone. Shear stresses were underestimated by about 47%, leading to an insufficient design. Concrete wall thickness was reported to be inadequate.

1.23.2   THUNDER HORSE PLATFORM

Another example is the accident at the Thunder Horse platform, shown in Figure 1.20. Thunder Horse production platform is in 1920 m of water in the Mississippi Canyon Block 778/822, about 150 miles (240 km) southeast of New Orleans. Construction costs were around U.S. $5 billion, and the platform is expected to operate for about 25 years. The hull section was constructed in 2004. In July 2005, Thunder Horse was evacuated due to the threat caused by Hurricane Dennis. After the hurricane had passed, the platform was inspected, and assessment reports did not mention any

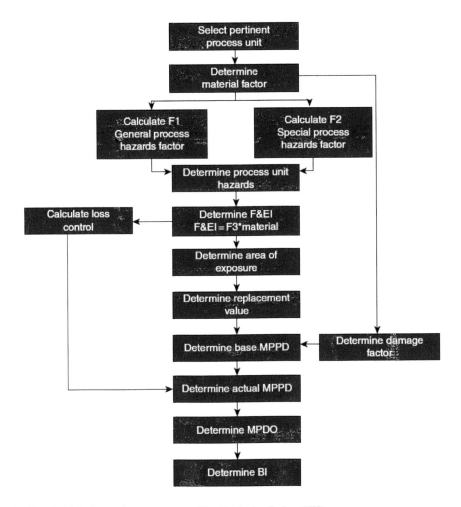

FIGURE 1.18 Flowchart to compute Fire Explosion Index (FEI).

TABLE 1.16

The Degree of Hazard for Different Dow FEI

Dow FEI	Degree of Hazards
1-60	Light
61-96	Moderate
97-127	Intermediate
128-158	Heavy
159 and above	Severe

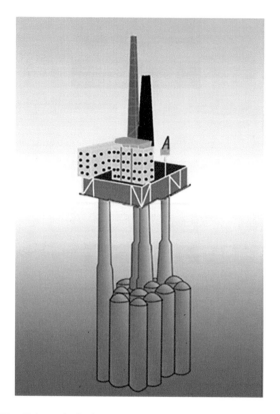

FIGURE 1.19   The Sleipner A platform.

FIGURE 1.20   Thunder Horse platform.

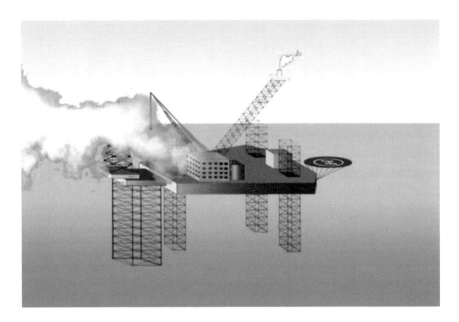

FIGURE 1.21   Timor Sea oil rig.

damages to the platform's hull. Interestingly, an incorrectly plumbed pipeline was allowed. Water flowed freely among the several ballast tanks, which initiated the platform to tip into the water. As a severe consequence of the accident, world oil prices increased because of speculation of oil shortage. The platform was subsequentially rehabilitated within a fortnight after Hurricane Dennis, and subsequent hits by Hurricane Katrina six weeks later did not damage the platform.

1.23.3   TIMOR SEA OIL RIG

Another example is the Timor Sea oil rig accident, shown in Figure 1.21. The leaking Timor Sea oil rig caught fire on November 2, 2009. While the oil spill resulted in severe environmental damage, the cause of the fire was not known immediately; personnel onboard were moved out safely without any fatal injuries.

1.23.4   BOMBAY HIGH NORTH IN OFFSHORE MUMBAI

Bombay High North's (BHN) massive platform in offshore Mumbai High Field was gutted in a devastating fire on July 27, 2005. In less than two hours, BHN was reduced to molten metal, as shown in Figure 1.22. The platform remained a beehive of activity for 24 years, halted due to the accident; it was later retrofitted and made functional.

From the events discussed, it is essential to know that the causes of failures are unknown in most cases. Even post-accident studies could not trace out the fundamental causes of the accident but hinted toward a set of complex reasons (Prem

FIGURE 1.22   Burnt out of Bombay High North platform.

et al., 2010). However, from an engineering perspective, one can understand that the causes are primarily due to oversight in the design stage or during operation/maintenance (ABS, 2004, 2014). As the consequences of such accidents severely impact the world's economy through oil pricing, it is imperative to note that risk analyses are becoming increasingly important to ensure that at least such events are not repeated (Erik et al., 2008; Chandrasekaran, 2016a, 2016b; Chandrasekaran and Jain, 2016; Mayerhofer et al., 2010; Micahilidou et al., 2012). It shows the importance or necessity of QRA tools (e.g., HAZOP) and their applicability to offshore platforms or any process industry in general at different stages: (i) front-end engineering design stage; (ii) fabrication, construction, and commissioning stage; and (iii) operational stage, etc. (Aktan et al., 1998a, 1998b; Chandrasekaran and Pannerselvam, 2009).

2 Health, Safety, and Environmental (HSE) Practices

Summary

This chapter deals with HSE practices commonly followed in the process industries. The chapter introduces HSE to the learner through a detailed glossary of terms while emphasizing the need for safety assurance in the process industries. A few case accident case studies are considered, highlighting the necessity of safety assessment and factors that can lead to accidents. The chapter also deals with detailed explanations of risk assessment methods and procedures. Logical risk analysis, a useful method for risk assessment and mitigation, is discussed with an illustrated example. Process industries possess potential hazards and operability problems, which need to be identified and resolved. Hazop methodology is presented in detail and explained with simple examples.

2.1 INTRODUCTION

Environmental, health, and safety (HSE) management is a crucial component of every organization and is especially important for managing operations in the oil and gas industry. The expectations of divisional conformity with the standard rules are typically considered when laying out HSE regulations. Since it is the most significant aspect of HSE covered by legislation in the last several decades, it serves as the foundation for HSE regulations in this era. It also establishes the framework for further regulations, laws, enforcement procedures, and employers' and others' broad responsibilities and obligations. To ensure the safety of both personnel and property, standards set by the HSE are confined to activities that are "reasonably practicable." General responsibilities for one's health and safety as well as the health and safety of teammates are mandated by HSE regulations, which also place general duties on employers to assist employees with minimal health and safety standards and members of the public.

2.2 GLOSSARY OF TERMS

ALARP: refers to reducing risk to "as low as reasonably practical" (ALARP) level. It involves balancing reduction in risk against time, trouble, complexity, and cost of achieving it. The cost of further reduction measures becomes unacceptably disproportionate to the additional risk reduction obtained.

Audit: It is a systematic, independent assessment to ascertain whether or not the HSE-MS and its operations adhere to the planned arrangement. The effectiveness of the system's implementation and suitability for achieving the company's HSE objectives and policies are also examined.

Client: An organization that contracts with independent contractors or subcontractors. The client in this contract is an oil and gas exploration company that will hire a contractor to carry out the work. Subcontractor(s) can subsequently get contracts from the contractor working as the client.

Contract: A contract that binds the parties and can be enforced in court or another appropriate forum, since it is enforceable against them. A contractor is an individual or business performing services for a client by a written or verbal contract. A subcontractor is a person or company performing part of the work stipulated within a contract and under contract to the original client or contractor.

Hazard: An object, physical consequence, or state that has the potential to harm people, the environment, or property.

HSE: Health, Safety, and Environment. These standards recognize that security and social responsibilities are integral elements of the HSE management system. HSE capability assessment is a method of screening prospective contractors to establish that they have the necessary experience, skills, and capability to handle the hazards involved in the work responsibly and have the knowledge and ability to complete it. The HSE Plan is comprehensive and lays out a particular contract's entire HSE management system, including any interface topics.

Incident: An event/incident or series of events that have caused or could have caused injury or illness to people and damage (loss) to the environment, assets, or third parties. It includes near-miss events also.

Inspection: A system of checking that an operating system is in place and is functioning satisfactorily. Usually, this is conducted by a manager and with prepared checklists. It is crucial to note that this is different from an audit.

Interface: A documented identification of relevant gaps (including roles, responsibilities, and actions) in each different HSE-MS of the participating parties in a contract, which, when added to the HSE plan, will combine to provide an operating system to manage all HSE aspects encountered in the contract to be handled as effectively and efficiently as possible.

Leading indicator: A measure that, if implemented, improves performance.

Third-party: Individuals, groups of people, or organizations other than the principal contracting parties that could be impacted by or involved with the contract.

Toolbox meeting: A meeting held by the employee workforce at the workplace to discuss HSE potential hazards encountered during work and the procedures in place to successfully manage these hazards. Usually, this is held at the start of the day's work, a process of continual awareness and improvement.

Accident: It describes the occurrence of a single or sequence of events that produce an unintentional loss. It refers to the occurrence of events only and not the magnitude of events.

Safety or loss prevention: The prevention of hazard occurrence (accidents) through proper hazard identification, assessment, and elimination of hazards.

Consequence: It is the measure of anticipated impact on the results of an incident.

Risk: It measures the magnitude of damage and its probability of occurrence. Stated unconventionally, it is the product of the chance that a specific undesired event will occur and the severity of the event's consequences.

Risk analysis: It is the quantitative assessment of risk using engineering evaluation and mathematical techniques. It involves estimating hazards, their probability of occurrence, and a combination of both.

Hazard analysis: It is the identification of undesirable occurrences that lead to the materialization of a hazard. It comprises an analysis of the mechanisms by which these undesired events could occur and estimating the extent, magnitude, and likelihood of harmful effects.

Safety program: A practical program identifies and eliminates existing safety hazards. An outstanding program prevents the existence of a hazard in the first place. Ingredients of a safety program are safety knowledge, safety experience, technical competence, safety management support, and commitment to safety.

There are two sets of regimes in implementing HSE practices, namely (i) goal-setting regime and (ii) rule-based regime. Goal-setting regime assigns the responsibility to a duty holder who evaluates the risk. He should demonstrate an understanding and control of the managerial concerns and issues related to the technical systems. He should keep pace with the emerging trend of safety practices and encourage workforce involvement. On the other hand, the rule-based regime consists of a legislator who sets the rules. He emphasizes strict compliance with the rules rather than auditing the outcomes. One of the main disadvantages is that the response in the latter method is too slow. It is interesting to note that both methods have merits and demerits. The former is considered to be employee-friendly while the latter is employer-friendly.

2.3 IMPORTANCE OF SAFETY

Risks are associated with every kind of work and workplace in daily life. Levels of risk involved in some industries may be higher or lower due to the consequences involved. These consequences affect industry and society, which may negatively impact the market depending upon the level of risk involved (Ale, 2002). Therefore, it is imperative to prevent death or injury to workers and the general public, prevent physical and financial loss to the plant, and prevent damage to the third party and the environment. Hence, rules and regulations for ensuring safety are framed and strictly enforced in offshore and petroleum industries, considered some of the most hazardous industries (Arshad, 2011). The prime goal is to protect the public, property, and environment in which they work and live. It is a commitment for all industries and other stakeholders toward the interests of customers, employees, and others. One of the primary objectives of the oil and gas industries is to carry out the intended operations without injuries or damage to equipment or the environment. Industries need to form rules, including all applicable laws and relevant industry standards of practice. Industries need to evaluate the HSE aspects of equipment and services continuously. The oil and gas industry needs to believe that effective HSE management will ensure good business. Continuous improvement in HSE management practices will yield good returns for the company, apart from guaranteeing the well-being of the employees (Bottelberghs, 2000). Every employee should feel responsible and accountable for HSE from top management through the entry level Industries must be committed to integrating HSE objectives into management systems at all levels. It will enhance the business and increase success by reducing risk and adding value to customer service.

2.3.1 WHAT IS SAFETY?

Safety is a healthy activity preventing exposure to hazardous situations. By remaining safe, the disastrous consequences are avoided, saving the lives of humans and plants in the industry. Safety is important because living creatures worldwide prefer to be safe rather than risk themselves to unfavorable circumstances. The term "safety" is always associated with risk. When the chances of risks are higher, the situation is said to be highly unsafe. Therefore, risk must be assessed and eliminated, and safety has to be assured.

Safety assurance is crucial in offshore and petroleum industries, as they are highly prone to hazardous situations. Two good reasons for practicing safety measures are: (i) investment in an offshore industry is several times higher than that of any other process/production industry across the world and (ii) offshore platform designs are very complex and innovative, and hence, it is not easy to reconstruct the design if any damage occurs (Bhattacharyya et al., 2010a, 2010b). Before analyzing the importance of safety in offshore industries, one should understand the main concerns in petroleum processing and production. Safety can be ensured by identifying and evaluating the hazards at every operation stage. Identification and assessment of risks at all stages of operation are essential for monitoring safety, both in quantitative and qualitative terms. The prime importance of safety is to prevent death or injury to workers in the plant and the public around it. Safety should also be checked regarding financial damage to the plant as investment is vaster in the oil and petroleum industries than in any other industry. Safety must be ensured so the surrounding atmosphere is not contaminated (Brazier and Greenwood, 1998).

Piper Alpha suffered an explosion in July 1988, which is still regarded as one of the worst offshore oil disasters in the history of the United Kingdom (Figure 2.1). About 165 persons lost their lives, along with 220 crew members. The accident is attributed primarily to human error and is a primary wake-up call for the offshore industry to review safety protocols. Estimation of property damage is about $1.4 billion. It is understood that the accident was mainly caused by negligence. Condensate leaked, and the accident occurred due to maintenance being carried out concurrently on one of the safety valves of the high-pressure condensate pumps. The condensate line was temporarily sealed with a blind flange after one of the gas condensate pumps' pressure safety valves was removed for repair. This work still needs to be finished during the day shift. Unaware that one of the pumps was undergoing maintenance during the previous shift, the night crew activated the backup pump. After that, other explosions occurred due to the blind flange—which had firewalls—unable to withstand the pressure. An explosion occurred due to the inability to stop the

FIGURE 2.1 Piper Alpha disaster.

gas flow from the Tartan Platform, intensifying the fire. The automatic firefighting equipment remained deactivated because the divers had worked underwater before the event. Thus, it may be deduced that human error and inadequate training in shift handovers were the causes of this catastrophic event. After this incident, significant (and stringent) changes were brought about in the offshore industry regarding safety management, regulation, and training (Kiran, 2014).

Exxon Valdez, carrying 180,000 tons of crude oil, was traveling from Valdez, Alaska, when it collided with an iceberg on March 23, 1989, puncturing 11 cargo tanks. Nineteen thousand tons of crude oil were lost in a few hours. Approximately 37,000 tons had been lost when the ship was refloated on April 5, 1989. Furthermore, roughly 6600 km2 of the nation's best fishing grounds and the shoreline around them were covered in oil. The government and industry found it challenging to salvage the situation due to the extent of the spill and its remote location. About 20% of the 18,000 tons of crude oil that the ship was carrying when it hit the reef was spilled (Figure 2.2). Safety plays a vital role in the offshore industry's global economy. Safety can be achieved by adopting and implementing control methods such as regular monitoring of temperature and pressure inside the plant using well-equipped coolant system, proper functioning of check valves and vent outs, practical casing or shielding of the system, and checking for oil spillages into the water bodies; by thoroughly ensuring proper control facilities, one can avoid or minimize the hazardous environment in the offshore industry (Chandrasekaran, 2011a, 2011b).

2.4 OBJECTIVES OF HSE

The main objective is to describe how clients can select suitable contractors and award contracts to improve the client and contractor management on HSE performance in

FIGURE 2.2 Oil spill in Exxon Valdez.

upstream activities. The document title does not include brevity, security, and social responsibilities; however, they are recognized as integral elements of the HSE management systems. Active and ongoing participation by the client, contractor, and subcontractors is essential to achieve the goal of effective HSE management. While each has a distinct role in ensuring the ongoing safety of all involved, there is an opportunity to enhance further the client–contractor relationship by clearly defining roles and responsibilities, establishing attainable objectives, and maintaining communication throughout the contract lifecycle.

HSE practice aims to improve performance by the following:

- Provide effective management of HSE in a contract environment so both the client and the contractor can devote their resources to improve HSE performance.
- Facilitating the interface of the contractor's activities with those of the client, other contractors, and subcontractors so that HSE becomes an integrated activity of all facets of the process.

The foregoing guidelines are formulated and provided to assist clients, contractors, and subcontractors in clarifying the process of managing HSE in contract operations (Chandrasekaran, 2014b, 2014c). This generated document differs from the professional judgment needed to recommend the specific contracting strategy. Each reader should analyze their situation and modify the information provided in this document to meet their needs and obtain appropriate technical support wherever required. The Oil and Gas Production Secretariat is the custodian of these guidelines and will initiate updates and modifications based on user review and feedback through periodic meetings. Generally, these guidelines are not intended to take precedence over a host country's legal or other requirements (Chandrasekaran, 2011e).

2.5   NEED FOR SAFETY

HSE guidelines provide a framework for developing and managing contracts in the offshore industry. While HSE aspects are essential in creating a contract strategy, these guidelines only cover some vital elements of the contract process. They prescribe various phases of the contracting process and associated responsibilities of the client, contractors, and sub-contractors. It begins with planning and ends with an evaluation of the contract process.

Employers establish teams, such as quality assurance (or control) teams, to involve employees in the quality process. Employees are empowered to stop an entire production line if they become aware of any problem affecting production or quality. This common industrial practice ensures increased participation in improving quality standards and reducing the cost line. A similar trend is necessary in practicing safety norms as well. Unfortunately, it is observed that, in many process industries, employees are not involved in the safety process, except that they are members of the safety committee. However, it is essential to realize that, if one desires to improve something for which employees are responsible, one should establish it as a critical component of their workday by making it an essential element of their business. Involving the employees in the safety assurance program gives them a keen sense of consciousness and ownership; results include better production and lower prices. Punishing a worker who broke a safety principle is not recommended, but the supervisor or manager who sanctioned the violation through his/her silence. The task of the supervisor or manager is to guarantee that the job is performed correctly and safely.

Managers should also be responsible for answering the perceived challenges as part of the system that challenges safety. Long-lasting safety success can only be assured if the management team is a function of the safety effort. The goal of every organization should be to build a safety culture through employee engagement. By getting employees involved in performing inspections, investigations, and other procedures, the needs of safety and health programs can be easily met. Employee safety can be maximized by creating a safety culture through increased consciousness. A skillful director of an oil company will make every effort to improve and regularize the outcome of the business in its entirety. However, it is common for a manager to excel in specific fields. In the workplace, several micro issues must be successfully managed for the company to succeed. One may establish quotas or reward individual achievements to recognize the outstanding production effort of a particular employee or a group of employees. Alternatively, one should ensure that safety is even unknowingly maintained during this rigorous task. As for safety and health, if the company tries to manage them for maximum success, it is also necessary to execute the program similarly. Safety managers are the experts who coordinate efforts and keep top management informed on issues linked to safety and health.

Policies and procedures, along with the signs and warnings, provide some restraint measures. The point of control is only as practical as the level of enforcement of the indemnities. Where enforcement is weak, control and, thus, compliance is weak as well. The best-suited example is the signboard, which is utilized to master the speed point of accumulation on highways. Nevertheless, only where the signs are strictly enforced can the drivers comply with the indicated speed limits.

In most cases, they will drive as fast as they think law enforcement will take into account. Therefore, it is not the signal that controls speed on the highway; it is the degree of enforcement established by local law. Thus, to prevent employee injury and sickness, one should maximize workplace safety and health management.

2.6   ORGANIZING SAFETY

Major accidents reported in oil and gas industries in the past are essential sources of information for understanding safety (Nivolianitou et al., 2006; Nordic Committee, 1977; Norwegian Petroleum Directorate, 1985). Lessons learned from these accidents, through detailed diagnosis, will help prevent similar accidents in the future. It is evident from the literature that, in the last 15 years, significant accidents in the offshore industry have declined (Khan and Abbasi, 1999). The essential experiences gained from these events are blanked out, and the information may not be brought forward to future generations if analyses of such accidents are not reported (OCS, 1980). The primary risk groups in the offshore and oil industry are blowouts, hydrocarbon leaks on installations, hydrocarbon leaks from pipelines/risers, and structural failures (Vinnem, 2007). Some of the major accidents that took place in the past and the lessons learned from these accidents are discussed in the next section.

2.6.1   EKOFISK B BLOWOUT

On April 23, 1977, a blowout occurred in the steel jacket wellhead platform during a workover on a production well. The Blow Out Preventer (BOP) was not in place and could not be reassembled on demand. All the personnel on board were rescued through the supply vessel without injuries, but the accident resulted in an oil spill of about 20,000 m3. The well was then mechanically capped seven days after the event, and production was shut down for half a dozen weeks to allow cleanup operations. Although the Ekofisk B blowout did not result in human death or material damage and was exclusively limited to spills, an important lesson learned is that capping a blowout is possible, although it requires time. It may be vital information from a design point of view, which can be considered in the modeling and analysis of BOPs (Kiran, 2012) (Figure 2.3).

2.6.2   ENCHOVA BLOWOUT

On August 16, 1984, a blowout occurred on the Brazilian fixed jacket platform, Enchova-1. It produced 40,000 barrels of oil and 1,500,000 m3 of gas daily through 10 wells. The first fire was due to ignition of gas released during drilling, which was under constraint. However, the fire due to oil leakage led to a knock. The ensuing flame was blown out late the following day. The platform's drilling equipment was gutted, but the remainder of the platform remained intact. Thirty-six people were killed while evacuating as the lifeboat malfunctioned, and 207 survivors were rescued from the platform through helicopters and lifeboats. The most vital lesson from the accident was using conventional lifeboats for evacuation. The failure of hooks in the lifeboat gained attention and led to later improvements in the design. Lack of competence to control the release mechanism led to stringent personnel training on safety operations during rescue and emergencies (Chandrasekaran, 2011d) (Figure 2.4).

FIGURE 2.3 Ekofisk blowout.

FIGURE 2.4 Enchova blowout

2.6.3 West Vanguard Gas Blowout

The semisubmersible drilling unit, West Vanguard, experienced a gas blowout on October 6, 1985, while conducting exploration drilling in Haltenbanken, Norway. During drilling, the drill bit entered a thin gas layer about 236 m below the sea bottom. It caused an influx of gas into the wellbore, which was followed by a second influx of gas after a day; the third influx of gas had a gas blowout. It was noticed that the drilling operation was carried out without using BOP. When the drilling crew realized the gas blowout happened, inexperienced personnel started pumping heavy mud and opened the valve to divert gas flow from the drilling stack. However, within minutes, erosion in the bends of the diverter caused the escape, and the gas entered the cellar deck from the bottom. An attempt to release the coupling of the wellhead of the marine riser, located on the seabed, was unsuccessful due to the ignition hazard in all areas of the platform. Ignition finally occurred from the engine room 20 minutes after the initial start of the event, which led to an intense explosion and a fire. Two lifeboats were launched for the crew members immediately after the burst. One of the lessons learned was the time management of launching lifeboats, which saved the lives of people onboard. However, inexperienced attempts to divert the gas flow away from the drilling stack remained a vital lesson to learn (Figure 2.5).

2.6.4 Ekofisk-A Riser Rupture

The riser of the steel jacket wellhead platform Ekofisk Alpha ruptured due to fatigue failure on November 1, 1975. The failure occurred due to insufficient protection in the splash zone and led to rapid corrosion. Leaks occurred at once at a lower part of the living quarters, causing an explosion and flame propagation. Intense flame remained for a short duration as the gas flow was immediately shut down; the blast was eliminated within two hours due to the efficient design of the firefighting system. Only a modest amount of damage to the platform was caused by the fire. The most

FIGURE 2.5   West Vanguard gas blowout.

important lesson learned from the accident is about the location of the riser below the living quarters (Chandrasekaran, 2010b). The best training and emergency evacuation procedures adopted and practiced by the crew resulted in minor injuries with no fatalities. The platform only suffered limited fire damage due to the short duration of intensive fire loads.

2.6.5   PIPER-A EXPLOSION AND FIRE

On July 6, 1988, an ignition caused a gas leak from the blind flange in the gas compression area of Piper A. The explosion load was estimated to be about 0.3–0.4 bar overpressure (Planas-Cuchi et al., 2004; Sutherland and Cooper, 1991). The first riser rupture occurred after 20 minutes, from which the fire increased dramatically; this resulted in further riser ruptures. The personnel who escaped the initial explosion gathered in the accommodation and were not given any further instructions about the escape and evacuation plans. Onboard communication became nonfunctional due to the initial stages of the accident. Evacuation with helicopters' aid was impossible due to blasts and smoke around the platform. A total of 166 crew members died in the incident. Most of the fatalities were due to smoke inhalation inside the accommodation, which subsequently collapsed into the ocean. From a design perspective, the accident could have been avoided by the location of the central room, radio room, radio room, and accommodation, which were very close to the gas compression area (Chandrasekaran, 2015a). Further, not protecting them from blasts and fire barriers was also a design fault. The location of accommodation on the upside of the installation led to the quick accumulation of smoke within the quarters, which is also a major design fault. Lessons learned from the operational aspects are as follows: The fire water pump was not kept on automatic standby for long. It was a severe installation failure, which led to the unavailability of water for cooling oil fire. Figure 2.6 shows a schematic representation of the Piper Alpha explosion.

FIGURE 2.6   Piper Alpha explosion.

From the accident cases discussed previously, it is understood that there is limited knowledge of forecasting the consequences of such incidents. Past experiences alone cannot calculate the outcomes' sequence (Kletz, 2003). It is because such accidents are very uncommon and cannot be predicted. However, in most cases, catastrophic consequences could have been avoided by taking proper care during the design stage and by imparting emergency evacuation training to all personnel on board.

2.7   RISK

Fatality and damages caused to the human and material property will result in a financial loss to the investor. Risk involves avoidance of loss and undesirable consequences. Risk involves probability and estimate of potential losses as well (TNO, 1999; Zhang and Aktan, 1995). According to ISO 2002, risk is the combination of an event's probability and outcome. ISO 13702 defines risk as the probability at which a specified hazardous event will occur and the harshness of the effects of the case. Mathematically, risk (R) can be expressed for each accident sequence, as follows:

$$R = \sum_i (p_i C_i) \qquad\qquad (2.1)$$

Where p_i is the probability of accidents and C_i is the consequences, while the summation is applied to a group of accidents considered to quantify the risk. The foregoing expression gives a statistical look to the risk definition, which often means that the value in practice shall never be discovered. If the accident rates are rare, an average value will have to be assumed over a long period, with low annual values. If, for 50 years, one has reported only about six significant accidents with ten fatalities, then this amounts to about 0.2 per year. Therefore, risk converts an experience into a mathematical term by attaching the consequences of the events. Risk is a post-evaluation of any event or incident, but the risk can also be predicted with appropriate statistical tools (Chandrasekaran and Kiran, 2014a, 2014b) (Figure 2.6).

2.8   SAFETY ASSURANCE AND ASSESSMENT

Safety and risk are contemporary. Safety is a subjective term, whereas risk is an abstract term. Safety cannot be quantified directly, but it is always addressed indirectly using risk estimates. Risk can be classified into individual risk and societal risk. Personal risk is the frequency at which an individual may be expected to sustain a given level of harm from realizing the hazard. It usually accounts only for the risk of death and is expressed as risk per year or Fatality Accident Rate (FAR). It is given by:

$$\text{Average individual risk} = \frac{Number\, of\, Fatalities}{Number\, of\, People\, at\, Risk} \qquad\qquad (2.2)$$

Societal risk is the relationship between the frequency and number of people suffering harm from any hazard. It is generally expressed as FN curves, which show the relationship between the cumulative frequency (F) and the number of fatalities (N). It can also be expressed in the annual fatality rate, in which the frequency and fatality data are combined into a single, convenient group measure. As it becomes essential to quantify risk, risk estimates are attractive only because of the consequences associated with the term. However, for the consequences, the risk remains a mere statistical number. Now, one is interested to know methods to estimate loss. This is because financial implications that arise from the consequences can be easily reflected in the company's balance sheet. Unfortunately, only one method is capable of measuring accident and loss statistics concerning all required aspects.

Three methods are used in the offshore industry to quantify risk. They are as follows:

1. Occupational Safety and Health Administration, US Department of Labor (OSHA).
2. Fatal Accident Rate (FAR).
3. Fatality rate or deaths per person per year.

The common factor is that all the methods report the number of accidents and fatalities for a fixed number of working hours during a specified period, which is unique and familiar among them (Chandrasekaran, 2015a).

2.9 LOGICAL RISK ANALYSIS

Frank and Morgan (1979) proposed a systematic method of financing risk and presented a scheme for risk reduction. Their model applies to any process industry and, therefore, is also valid for oil and gas industries. Before using this method for targeting risk reduction, the whole company is subdivided into several departments. This division can be based on functional or administrative aspects. This method involves six steps of risk analysis, which are as follows:

Step 1: Compute the risk index for each department.

Each department inherently has a risk level that needs to be identified first. It can be done by evaluating the hazards present and the control measures available. It is also called the first level of risk assessment. It is computed by preparing a checklist, as shown in Table 2.1. Control and hazard scores for all the departments are established from the checklist, as shown in Table 2.2.

The hazard checklist has six groups of hazards. There are scores associated with each hazard within each group. These scores are summed up for hazards applied within that group. The hazard score for a group is given by:

$$\text{Hazard score} = Sum \times Hazard\,Weightage \tag{2.3}$$

The hazard score for each department is the sum of the scores computed for each of the six groups. Similarly, one can estimate the control scores as well. The control score for each department is the sum of the scores of each of the six groups, as tabulated previously. The control score for a group is given by:

$$\text{Control score} = Sum \times \text{control measure } Weightage \tag{2.4}$$

The risk index can be calculated as follows after determining each department's hazard and control scores. The risk index may be positive or negative depending on each department's control measures and hazard groups.

$$\text{Risk index} = \text{Control score} - \text{Hazard score} \tag{2.5}$$

TABLE 2.1

Hazard Groups and Hazard Score

Rating points Hazard group and hazard (Group hazard factor in parentheses)

Fire/explosion potential (10)

2   Large inventory of flammables
2   Flammables are generally distributed in the department rather than localized
2   Flammables normally in the vapor phase rather than the liquid phase
2   Systems opened routinely, allowing flammable/air mix, versus a closed system
1   Flammables having low flash points and high sensitivity
1   Flammables heated and processed above flash point

Complexity of process (8)

2   Need for precise reactant addition and control
2   Considerable instrumentation requiring unique operator understanding.
2   Troubleshooting by supervisor rather than operator
1   Large number of operations and equipment monitored by one operator
1   Complex layout of equipment and many control stations
1   Difficult to startup or shutdown operations
1   Many critical operations to be maintained

Stability of process (7)

3   Severity of uncontrolled situation
2   Materials sensitive to air, shock, heat, water, or other natural contaminants.
2   Potential exists for uncontrolled reactions
1   Raw materials and finished goods that require special storage attention
1   Thermally unstable Intermediate
1   Obnoxious gases present or stored under pressure

Operating pressure involved (6)

3   Process pressure over 110 lb/in2 (gauge)
2   Process pressure above the atmosphere but less than 110 lb/in2 (gauge)
1   Process pressure ranges from vacuum to atmospheric
3   Pressures are process rather than utility-related
2   High-pressure situations are in operator
1   Excessive sight glass application
1   Non-metallic materials of construction in pressure service

Personnel/environment hazard potential (4)

3   Exposure to process materials poses high potential for severe burn or severe health risks
2   Process materials corrosive to equipment
2   Potential for excursion above Threshold Limit Value (TLV)
1   Spills and flumes have a high impact on equipment, people, or services
1   High noise levels make communication difficult

High temperatures (2)

1   Equipment temperatures exist in <100°C range (low)
2   Equipment temperatures exist in the 100 < 170°C range
3   Equipment temperatures exist in the 170 < 230°C range
2   High-temperature situations are m operator-frequented area
2   Overflows or leaks are fairly common
2   Heat stress possibilities from the nature of work or ambient air

TABLE 2.2

Control Scores and Control Group

Rating points control group control (group control factor in parentheses)

Fire protection (10)

4   Automatic sprinkler systems capable of meeting demands

2   Supervisors and operators are trained properly

1   Adequate distribution of fire extinguishers

1   Fire protection system inspected and tested with regular frequency

1   Building and equipment provided with the capability to isolate and control fire

1   Special fire detection and protection provided where indicated

Electrical integrity (8)

3   Electrical equipment installed to meet National Electrical Code area classification

1   Electrical switch labeled to identify equipment served

1   Integrity of installed electrical equipment maintained

1   Class I, Division 2 installations provided with sealed devices Explosion-proof equipment provided or purged reliably and sound electrical isolation between hazardous and non-hazardous areas.

1   All electrical equipment capable of being locked out

1   Disconnects provided, identified, inspected, and tested regularly

1   Lighting securely installed and facilities properly grounded

Safety devices (7)

3   Relief devices are provided, and relieving is to a safe area

2   Confidence that interlocks and alarms are operable

2   Operating instructions are complete and current, and the department has continued training and retaining the program

1   Safety devices are appropriately selected to match the application

1   Critical safety devices identified and included in the regular testing program

1   Fail-safe instrumentation provided

Inerting and dip piping (5)

2   Vessels handling flammables are provided with dip pipes

2   Vessels handling flammables provided with a reliable inerting system

2   Effectiveness of inerting assured by regular inspection and testing

1   Inerting instruction provided and understood

1   Inerting system designed to cover routine and emergency startup

1   Equipment ground visible and tested regularly

1   Friction hot spots identified and monitored

Ventilation/Open construction (4)

3   No flammables exist, or open-air construction is provided

2   Local ventilation is provided to prevent unsafe levels of flammable, toxic, or obnoxious vapors

2   Provision made for containing and controlling large spills and leaks of hazardous materials

1   Building design provides for natural ventilation to prevent the accumulation of dangerous vapors

1   Sumps, pits, etc., non-existent or else properly ventilated or monitored

1   Equipment entry is prohibited until a safe atmosphere is assured

(Continued)

TABLE 2.2 (Continued)
Control Scores and Control Group

Rating points control group control (group control factor in parentheses)
Accessibility and separation (2)

2   Critical shutdown devices and switches visible and accessible
2   Adjacent operations or services protected from exposure resulting from an incident in the concerned facility
2   Operating personnel protected from hazards by location
1   Orderly spacing of equipment and materials within the concerned facility
1   Adjacent operations offer no hazard or exposure
1   Hazardous operations within the facility well isolated

Step 2: Determine the relative risk for each department.

The aim is to rank the departments and not the individual hazards present in the plant. The department with the highest risk index (highest positive value) will likely need little hazard reduction. A high-risk index means that the controls are adequate. Those departments will require less funds than other departments to mitigate/eliminate/reduce hazards. Use the best department risk score as the base reference. All curves are normalized concerning the best department. It is done by subtracting the risk score of the best department from the risk scores of the concerned department. This adjustment will result in the relative risk of the best department being zero.

Step 3: Compute the percentage risk index for each department.

It indicates each department's relative contribution to the plant's total risk. The relative risk of each department is converted to a percent of the total risk by a simple procedure. The total risk of all departments is the sum of the absolute value of the relative risk of each department. The percent risk index is given by:

$$\% \ Risk \ Index = \frac{Relative \ Risk_i}{\sum_0^i Relative \ Risk} \times 100 \qquad (2.6)$$

Step 4: Determine composite exposure dollars for each department.

The estimated risk is subsequently converted to financial value now. It estimates the financial value of risk for each department. Composite exposure dollars are the sum of the monetary value of three components: property value, (ii) business interruption, and (iii) personnel exposure. Property value is estimated by the department's replacement cost of all materials and equipment at risk. Business interruption is computed as the product of the unit cost of goods and production per year and expected percentage capacity. Personnel exposure is the product of the total number of people in the department during the most populated shift and each person's monetary value.

Step 5: Compute composite risk for each department.

For each department, the composite risk is the product of composite exposure dollars and the percentage risk index of that department. This value represents the relative risk of each department. Units for composite risk are in dollars. The composite risk for each department is given by:

$$\text{Composite risk} = (\text{composite exposure}) \times (\% \text{ Risk Index}) \qquad (2.7)$$

Step 6: Risk ranking.

It is the final step in the process. Risk ranking of the departments is done based on the composite risk as this will help the risk managers decide the fund requirement for each department to mitigate risk or at least control risk. Departments should be ranked from the highest composite score to the lowest.

Example problem

Now, let us consider an example to understand the application of Frank and Morgan's risk analysis. Relevant data for each department is given in Table 2.3.

From the given input data, the risk index is calculated using Eq. (2.5). For example, the risk index of department A is given as (304 - 257 = +47); a positive value indicates that the department is exercising better control in risk management. The risk index of all departments is computed similarly. Results are tabulated in Table 2.4. The entries in the table are self-explanatory.

As seen in the foregoing table, the composite risk is highest for Department A. As per Morgan's analysis, the budget allotted for safety is distributed among the departments according to the risk ranking. It implies that more money is required to control risk and initiate risk control measures in Department A. It is interesting to note that Department E has a higher risk index than Department C, but the risk ranking is reversed. It is because of the composite cost on the risk index. Hence, this method helps to analyze the safety requirements of the departments comprehensively, including the composite exposure of each department. Computations of risk rankings for other departments are shown in Table 2.4. The goal is to reduce the potential losses within the plant while identifying the crucial department responsible for higher risk. This method also helps safety executives pay attention to those essential departments. Morgan's method is one of the best-employed tools for such problems, as seen in the literature, and possibly the easiest method to attempt financing risks (David et al., 2007).

2.10   DEFEATING ACCIDENT PROCESS

Different steps involved in an accident include initiation, propagation, and termination. Initiation is the event that starts the accident. It should be reduced to avoid a significant accident. The procedures to control the initiation of the events are grounding, inerting, maintenance, improved design, and training to reduce human error. Propagation is the event that expands the accidents. These events should be curtailed effectively. Some procedures to control the propagation include emergency material transfer, fewer inventories of chemicals, use of nonflammable construction materials, and installation of emergency and shutdown installation valves (Amenola et al., 1992). Termination is the event that stops the accident. It should be increased to have better control over the accident. Some of the procedures to control termination are end-of-pipe control measures, firefighting equipment, relief systems, and sprinkler systems.

TABLE 2.3
Data for Each Department of the Process Plant

| Exposure Dept. | Hazard Score | Control Score | Cost (x 10³) $ | | | Composite Score (x 10³) $ |
			Property Value	Business Interruption	Personnel	
A	257	304	2900	1400	900	5200
B	71	239	890	1200	653	2743
C	181	180	1700	720	1610	4030
D	152	156	290	418	642	1350
E	156	142	520	890	460	1870
F	113	336	2910	3100	1860	7870

TABLE 2.4
Risk Ranking Estimate

| Exposure Dept. | Risk Index | Relative Risk | % Risk Index | Composite Cost (x 10³) $ | | Risk Rank |
				Exposure	Risk	
A	47	−176	$\dfrac{-176}{-911} = 19.31$	5200	$\left(\dfrac{19.31}{100}\right)5200 = 1004$	1
B	168	−55	6.04	2743	166	5
C	−1	−224	24.59	4030	991	2
D	4	−219	24.04	1350	325	4
E	−14	−237	26.02	1870	487	3
F	223	0	0	7870	0	6
Check Sum		−911	100%			

2.11   ACCEPTABLE RISK

In offshore industries, risk cannot be avoided. Drilling, exploration, and production processes cannot be zero-risk zones as they have inherent factors that may lead to an unforeseen incident. Depending upon the environmental conditions prevailing, they can become an accident (Aven and Vinnem, 2007). It is, therefore, vital to understand that risk is accepted in offshore industries up to a certain level. According to the regulatory norms, the risk is acceptable and permissible in offshore industries. According to the United States Environmental Protection Agency, the risk of one in a million is sufficient for carcinogens. For non-carcinogens, the acceptable risk is a hazard index of less than one. According to the United Kingdom Health and Safety Executive, an acceptable FAR is unity. It is also interesting to note that even nonindustrial activities, which are part of daily routine, have risk indicators. Fatality statistics for everyday nonindustrial activities are given in Table 2.5.

TABLE 2.5
Fatality Statistics for Nonindustrial Activities (Lees, 1996)

Activities	FAR (deaths/10^8 hrs)
Staying at home	3
Traveling by car	57
Traveling by cycle	96
Traveling by air	240
Traveling by motorcycle	660
Rock climbing	4000

2.12 RISK ASSESSMENT

Risk assessment is the quantitative or qualitative value of risk related to a situation and a recognized hazard. Quantitative risk assessment involves estimating both the magnitude of potential loss and the probability of occurrence of that potential loss. Therefore, risk assessment consists of two stages: (i) risk determination and (ii) risk evaluation. Risk determination deals with numbers; hence, it is a quantitative approach. Risk evaluation deals with the events; hence, it is a qualitative approach. Risk is identified by continuously observing changes in risk parameters on the existing process and, therefore, a continuous process. Risk estimation is done by determining the probability of occurrences and the magnitude of consequences, which is the post-processing of the data identified during the former stage (Llyod's Register, 2005). Risk evaluation consists of risk aversion and risk acceptance. Risk aversion is determined by the degree of risk reduction and risk avoidance. Risk acceptance is the establishment of risk references and risk referents. Risk references are used to compare values, and risk referents are standards by which risk parameters are compared. For example, let us take a specific case for risk assessment of a chemical process plant. The National Academy of Sciences identified four chemical risk assessment steps: hazard identification, dose-response assessment, exposure assessment, and risk characterization.

2.12.1 HAZARD IDENTIFICATION

The hazard identification process includes engineering fault assessment. It is used to evaluate the reliability of specific segments of a process plant in operation. It determines the probabilistic results (Moses and Stevenson, 1970; Moses, 1977). The method employed in hazard identification is fault tree analysis.

2.12.2 DOSE-RESPONSE ASSESSMENT

It describes the quantitative relationship between the amount of exposure and the extent of toxic injury. The hazardous nature of various materials needs to be assessed before their effects are estimated. The dose-response assessment outcome is a linear equation relating to exposure to the disease, which is obtained by the regression analysis of the data.

2.12.3 EXPOSURE ASSESSMENT

It describes the nature and size of the population exposed to the dosing agent, its magnitude, and the duration of exposure. This assessment includes the analysis of toxicants in air, water, or food.

2.12.4 RISK CHARACTERIZATION

Risk characterization is the integration of data and analysis. It determines whether the person working in the process industry and the public in the nearby vicinity will experience the effects of exposure. It includes estimating uncertainties associated with the entire process of risk assessment.

2.13 APPLICATION ISSUES IN RISK ASSESSMENT

Risk assessment often relies on inadequate scientific information or lack of data. For example, any data related to repair may not be helpful to assess newly designed equipment. It means that, even though the data available is less, all data related to that event cannot be considered qualified data to do risk assessment. In toxicological risk assessment, the data associated with using them in animals is irrelevant to predicting their effects on humans. Therefore, to do a risk assessment, one uses probabilistic tools for which data size is one of the main issues. Due to the limited data available in terms of occurrences of events (as accidents are fewer) and their consequences, risk analysts use a conservative approach. They end up overestimating the risk by using a statistical approach. Alternatively, one can also estimate risk on a comparative scale. The conservative approach is a quantitative risk assessment, which identifies the frequency of an event and its severity. After placing the frequency and severity, risk rankings are determined to identify the critical events. Attention is paid to risk reduction or mitigation of these events instead of examining the whole process repeatedly. It is seen as one of the practical tools of risk reduction. The comparison technique is a qualitative risk assessment done by conducting surveys and preparing a series of questionnaires. Based on the survey results, risk ranking is done.

2.14 HAZARD CLASSIFICATION AND ASSESSMENT

The first step in all risk assessment, or Quantitative Risk Assessment (QRA), studies hazard identification (HAZID). HAZID aims to identify all hazards associated with planned operations or activities (Chandrasekaran et al., 2010). It provides an overview of risk, which helps plan further risk assessment analysis. It provides an overview of different types of accidents in the industry, ensuring no significant hazards are overlooked. Some of the terminologies commonly used in hazard classification and assessment are discussed in the following paragraphs:

Hazard refers to a chemical or physical condition that has the potential to cause damage to people, property, or the environment. A hazard is a scenario, which is a situation resulting in more likelihood of an incident. *Incident* refers to a loss of or contamination of material or energy. All incidents do not propagate to accidents. *Risk* is the realization of a hazard. The incident becomes an accident. *Hazard analysis* is identifying undesired events that lead to materializing a hazard. It includes an analysis of the mechanisms by which these undesired events could occur. It also includes estimating the extent, magnitude, and likelihood of harmful effects.

2.15 HAZARD IDENTIFICATION

Hazard identification deals with the engineering failure assessment. It evaluates the reliability of specific segments of a plant in operation to determine the probabilistic results of its operational and design failure. Fault tree analysis is one of the common forms of engineering failure assessment. Hazards common in oil and gas industries are only identified once an accident occurs. It is, therefore, essential to recognize the dangers to reduce risk. Some frequently asked questions that lead to hazard identification are as follows:

  (i) What are hazards?
 (ii) What can go wrong and how?
(iii) What are the chances that they can go wrong?
 (iv) What are the consequences if they go wrong?

The answer to the first question can be obtained by doing HAZID. The answer to the question of what can go wrong and how can be obtained by doing a risk assessment will subsequently lead to assessing the probability of failure. Answering questions (iii) and (iv) will lead to a detailed risk assessment. It is essential to document all the accidents and near-miss events occurring in the offshore industries to have a more comprehensive database. It helps estimate the frequency of occurrence of such accidents through detailed mathematical modeling with a higher accuracy. Documenting the accidents also simultaneously identifies consequences, which helps in risk assessment. Hazard evaluation combines HAZID and risk assessment; a flowchart is given in Figure 2.7.

Hazard evaluation can be performed at any stage of operation. It can also be performed during the preliminary stages of analysis and design of the process plant. During the initial design stages, hazard evaluation is done using Failure Mode Effective Analysis (FMEA), whereas, during the ongoing operation stages, it is done using Hazard and Operability Study (HAZOP). If the hazard evaluation shows low probability and minimum consequences, the system is called a *gold-plated system*. Such systems are examples of implementing potentially unnecessary and expensive safety equipment (David and William, 2007; HSE, 2010). As shown in Figure 2.7,

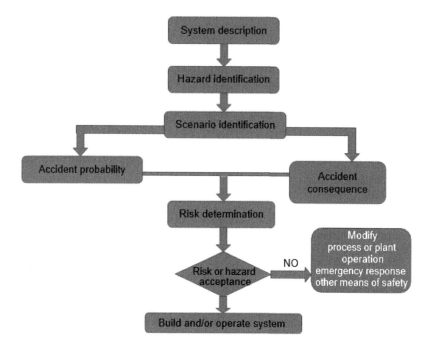

FIGURE 2.7   Flowchart for hazard evaluation.

the layout of hazard evaluation, the most critical step is risk acceptance. It is also complex because the level of risk acceptance is subjective to each organization and, hence, should be predefined; fortunately, oil industries follow such practices internationally to determine the level of risk acceptance (OISD-169, 2011; OISD-116, 2002; OSID-144, 2005; OISD-150, 2013; OGP, 2010).

Potentially unnecessary and expensive safety equipment and procedures are implemented in the system. One of the essential steps in hazard evaluation is to decide on the risk acceptance criteria. It is complex as the level of risk acceptance in oil and gas industries is subjective to each organization as well as the process and methods they adopt for exploration and production. Therefore, they should be predefined even before one attempts to perform a hazard evaluation. However, there are also standard procedures to define or determine risk acceptance levels.

2.15.1   IDENTIFICATION METHODS

The hazard identification methods are classified into various units as discussed as follows:

- **Process hazard checklists**: Refers to a list of items and possible problems in the process that must be checked periodically.
- **Hazard surveys**: Refers to the inventory of hazardous materials.

- *HAZOP*: Refers to Hazard and Operability Studies, which generally identifies the possible hazards present in any given process plant.
- *Safety review*: Refers to a less formal type of HAZOP study. The result depends upon the experience of the person conducting the review; hence, the review's outcome can be highly subjective.
- *What-if analysis*: This less formal method applies *what-if logic* to several investigations. For example, the question would be, *what if the power stops*? Answers to such questions yield a list of potential consequences and solutions.
- *Human error analysis*: Refers to a method used to identify parts and procedures of a process system. It is generally applied to the process with a higher probability of human error. An example shall be a fire alarm/buzzer system in the control panel.
- *Failure Mode, Effects, and Criticality Analysis (FMECA):* This method tabulates the equipment list and possible mechanical failure under working conditions. This study can identify the potential failure modes of each component present in the system and their effects of failure on the overall performance of the processing system.

2.16 HAZARD IDENTIFICATION DURING OPERATION (HAZOP)

Hazards arise due to deviations from standard processes. Deviations from the design intent always exist. It applies to the existing and new process plants. The primary purpose of the HAZOP study is to identify the potential hazards and the relative operability problems that arise due to the perceived deviations. HAZOP analysis identifies all possible risks and operational issues, recommends changes, and identifies areas that require further detailed studies. An up-to-date Process Flow Diagram (PFD) is required for a HAZOP analysis. It also requires a Process and Instrumentation Diagram (P&IDs), detailed equipment specifications, details of materials, and mass and energy balances. A team of experts experienced in a similar plant, and the technical and professionals conduct a HAZOP study.

2.16.1 HAZOP OBJECTIVES

HAZOP studies are carried out to identify the following; namely, i) any perceived deviations from the intended design/operation, ii) causes for those deviations, iii) consequences of those perceived deviations, iv) safeguards to prevent the causes and mitigate consequences of the perceived deviations, and v) recommendations in the design and operation to improve the safety and operability of the plant.

2.17 APPLICATION AREAS OF HAZOP

HAZOP is primarily used in chemical industries to estimate hazards that arise during operations; one such example can be seen in hazard studies carried out in the Flixborough disaster in 1974. It is a chemical plant in the United Kingdom that manufactures caprolactam, which is required to manufacture nylon. This incident

FIGURE 2.8   Explosion at rocket-fuel plant located in Nevada, Las Vegas.

occurred due to the rupturing of a temporary by-pass pipeline carrying cyclohexane at 150°C, which leaked and set into a fire. Within a few minutes after the fire initiation, about 20% of the plant's inventory got burnt, resulting in the spread of a vapor cloud over a diameter of about 200 m. It further resulted in an explosion of a nearby hydrogen production plant, which showed a cascading effect of the consequences.

Another similar example where HAZOP studies were applied successfully was to study the consequences of an explosion at a Rocket-fuel plant located in Nevada, Las Vegas, United States, as shown in Figure 2.8. The plant was destroyed a few seconds after the initiation of the explosion. The windstorm destroyed the roof structure and the glass. The fire was caused essentially due to the use of a welding torch in the windward direction. Studies reported that using HAZOP is seen to help predict the hazardous nature of the chemical release and its consequences.

2.18   ADVANTAGES OF HAZOP

HAZOP supplements the design ideas with imaginative anticipation of deviations. These may be due to equipment malfunction or operational error. In the design of new plants, designers sometimes overlook a few issues related to safety in the beginning. It may result in a few things that need to be corrected. HAZOP highlights these errors. HAZOP is an opportunity to correct these errors before such changes become too expensive or impossible. HAZOP methodology is widely used to aid loss prevention. HAZOP is a preferred tool for risk evaluation for a few reasons: (i) it is easy to learn; (ii) it can be easily adapted to almost all operations in the process industries; (iii) it is a standard method in contamination problems rather than chemical exposure or explosions; and (iv) requires no level of academic qualification to perform HAZOP studies.

HAZOP studies thoroughly examine the full description of the process. It systematically questions every part to establish the perceived deviations from the design intent. Once identified, an assessment is made to estimate the consequences of such deviations. If considered necessary, action is taken to rectify the situation. Though the method is imaginative, it is still systematic. It is more than a checklist type of review. It encourages the team to identify possible deviations and helps trace them all under operational conditions. HAZOP penetrates a greater depth of risk analysis of any process plant. HAZOP, when applied to the same type of plant, repeatedly improves safety, which is essential. Potential failures not noticed in the earlier studies can be easily highlighted using HAZOP.

2.19   STEPS IN HAZOP

Step 1: Define the design intent.

Defining the design intent is the first step in a HAZOP study. Let us explain the design intent using some examples. Consider the following:

1. Suppose a plant is in operation and must produce specific tons of chemicals annually.
2. An automobile unit must manufacture a certain number of cars yearly.
3. A plant has to process and dispose of a specific volume of effluent per year.
4. An offshore plant must produce specific barrels of oil annually.

The equipment is designed and commissioned to achieve the desired production capacity. To do so, each item, like the equipment, pump, and length of pipework, will need to be consistently functional in a particular (desired) manner. It is the design intent for that specific item and not the machinery or production capacity.

Step 2: Identify the deviations.

To understand the deviation in design intent, let us consider another example. A plant requires continuous circulation of cooling water at temperatures $x°C$ and xxx L/h. Heat exchanger does the cooling of the process. The effective working of the heat exchanger is mandatory for the plant. The design intent is the effective working of the heat exchanger. If the water supplied for circulation becomes greater than $x°C$, this will affect the production; hence, the deviation. Note the difference between the deviation and its cause. For example, failure of the pump would be a cause and not a deviation.

2.19.1   BACKBONE OF HAZOP

The backbone of HAZOP studies is the keywords used in the study. There are two types of keywords; namely, primary and secondary. The "primary" keyword focuses on a particular aspect of the design intent or an associated process condition. The "secondary" keywords suggest possible (perceived) deviations from the design intent. Combined with the primary keywords, they intend the required meaning. As HAZOP revolves around effectively using these keywords, it is necessary to understand their meaning and usage. Primary keywords reflect the intent of the process design and the operational aspects of the plant. A few examples are FLOW,

TEMPERATURE, PRESSURE, LEVEL, SEPARATE, COMPOSITION, REACT, MIX, REDUCE, ABSORB, CORRODE, ERODE. These keywords may sometimes need to be clarified. For example, let us take the word CORRODE. One may assume that the intention is that corrosion should occur as it refers to the design intent. Most of the plants are designed with the intent that corrosion should not occur during their lifespan, or if expected, they should be at most a specific rate. An increased corrosion rate would result in a deviation from the design. Hence, this keyword shall be used to indicate the deviation from the design.

Secondary keywords are applied in conjunction with the primary to suggest potential deviations. Examples: NO, LESS, MORE, REVERSE, ALSO, OTHER, FLUCTUATION, EARLY, etc. They convey the meaning of deviation from the design intent. For example, Flow/No indicates that there is no desired flow, which is a deviation from the design intent of FLOW. Another example could be the operational aspect, such as isolate/no. It should be noted that not all combinations of primary/secondary keywords are appropriate. For example, Temperature/No and pressure/reverse could be meaningless. Results of the HAZOP study are recorded in a desired format, termed a HAZOP report, as shown in Table 2.6.

Example problem

Let us consider an example problem of a flow line shown in Figure 2.9. FLOW/NO is applied to describe the deviation from the design intent. One of the reasons for the lack of flow could be the blockage of the strainer S1 due to the impurities present in the dosing tank T1. Consequences arising from the loss of dosing are incomplete separation in V1; additional causes may be cavitation in pump P1, which may result in damage if prolonged. While recording consequences, one should be explicit. For example, instead of recording as "No dosing chemical to the mixer," it is better to add a detailed explanation and the reason for no dosing chemical to the mixture. When assessing the consequences, one should not consider any protective systems or instruments already included in the design. Let us consider a case where the HAZOP team identified a cause for FLOW/NO in a system as spurious closure of an actuated valve. It is noticed that the valve position is displayed in the central control room, and an alarm exists in the control panel, indicating spurious closure of the valve. Even in this situation, one may consider adding the details in assessing the consequences and recommending a few additional control measures as safeguards against the identified cause. In the example under consideration, as the spurious closure of the valve could

---

TABLE 2.6

HAZOP Report Format

Description	Deviation	Cause	Consequence	Safeguard	Recommended Action

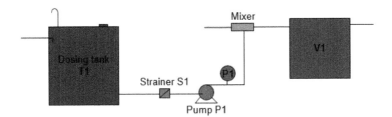

FIGURE 2.9　Example problem for Hazop.

increase pressure in the upstream line, which can lead to other cascading consequences like fire, etc., it is better to add additional safeguards despite the presence of an alarm system in the control room. Hence, while recording HAZOP reports, one should refrain from taking credit for the existing protective systems or instruments already included in the design but recommend additional/alternative safeguards.

Any protective devices that prevent the cause or safeguard against the adverse consequences should also be recorded in the HAZOP report. Safeguards need not be restricted only to hardware; one can also recommend periodic inspections of the plants as safeguard measures. If a credible cause results in a negative consequence, it must be decided whether some action should be taken along with its priority. If it is felt that the existing protective measures are adequate, then no action needs to be recommended in the report.

Recording of action falls into two groups: (i) action that removes/mitigates the cause or (ii) that eliminates the consequences. Recommended actions that address the consequences are more (the latter) as this directly impacts the cost control toward risk reduction. Generally, the former type is preferred over the latter, but it is only sometimes possible when dealing with equipment malfunction. One of the probable actions that could be recommended for the present example is to provide a strainer on the road tanker itself, which can restrict the entry of impurities to the tanker T1. However, one should be careful as such recommendations may result in choking the pump at the inlet section.

While recommending actions in the HAZOP report, one should not always recommend engineered solutions such as adding additional instruments, alarms, trip-off switches, etc. This is because any failure of mechanical systems does not resolve the actual hazard identified in the original process layout. Regarding the reliability of such devices, one should remember that their potential for spurious operation will cause unnecessary downtime. In addition, this may also result in increased operational costs in terms of maintenance, regular calibration, etc. Further complications arise if trained personnel are not appointed to operate the sophisticated protective systems; their maintenance is also equally complicated and expensive.

2.20　HAZOP FLOWCHART

HAZOP studies are not carried out on the whole layout of the process plant, but only on the chosen segments. Such segments are identified through preliminary studies

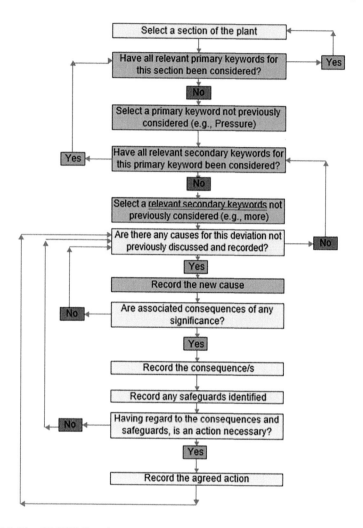

FIGURE 2.10   HAZOP flowchart.

such as HAZID (hazard identification). HAZOP procedure is discussed in the flow-chart given in Figure 2.10.

2.20.1   FULL AND EXCEPTION RECORDING

HAZOP reports prepared some years ago contained a partial recording of the poten-tial deviations and the associated consequences. Some of the negative consequences were also found to be recorded as they were helpful in the company's internal audit. This method of recording reduces time and effort, since they are handwritten records. Such methodology is called recording by exception. In this method, it is assumed that anything not included is deemed satisfactory.

On the other hand, recent practices have required reporting everything in detail. Each keyword is clearly stated as applied to the system under study. Even statements like "no cause could be identified" or "no consequence arose from the cause recorded" are also seen in these statements. It is called full recording. Complete recording reports verify that a rigorous study has been undertaken, as evident from the comprehensive document. It can assist in the speedy assessment of the safety and operability of modifications carried out later in the plant. With computer methods in practice, full recording has become more common. However, using a few MACRO words reduces the reading time of such total records. For example, MACRO words like "no potential causes identified," "no significant negative consequences identified," "no action required," etc., can be suitable for many studies that are carried out as a part of routine maintenance.

2.20.2 Pseudo-Secondary Words

A pseudo-secondary word is used along with the primary keyword when no appropriate secondary keyword is found suitable. For example, let us consider FLOW as one of the primary words for the report. Some combinations have credible causes, such as FLOW/NO, FLOW/REVERSE, etc., and a few combinations have no causes, such as FLOW/LESS, FLOW/MORE, FLOW/OTHER, etc. So, FLOW/REMAINDER can be used as a MACRO word that substitutes the meaning of a group of negations, as shown in the later set. Some of the pseudo-secondary words are ALL, REMAINDER, etc. After exploring all possible combinations of primary/secondary keywords, if no potential deviations could be identified, then FLOW/ALL can also be used in the report. Using pseudo keywords improves readability, eliminating countless repetitive entries in the report. However, the HAZOP report should initially mention a list of secondary keywords; otherwise, pseudo keywords may have ambiguous meanings.

2.21 WHEN TO DO HAZOP?

HAZOP studies are generally carried out to identify potential hazards and operability problems caused by deviations that arise from the design intent. If significant deviations are made during any recent modifications made in the process line, the changes should be verified for their safety through HAZOP studies. As a general practice in the oil and gas industries, HAZOP studies are carried out at periodic intervals of not later than six months. HAZOP studies should preferably be carried out as early in the design phase because this influences the changes in the design if deemed fit. Unfortunately, a good HAZOP study can only be done when a complete design is available. As a compromise, HAZOP is usually carried out as a final check when the detailed design is completed. HAZOP studies may also be conducted on an existing facility to identify the modifications that should be implemented to reduce risk and operability problems. The following situations generally necessitate HAZOP studies:

- When the design and detailed drawings are available at the initial concept stage.
- When the final P&ID is available.

- During the construction and installation stages, during which a set of valid recommendations are to be implemented.
- During the commissioning of the plant.
- During the plant operation ensure that the emergency and operating procedures are regularly reviewed and updated as required by OSID norms.

2.22   TYPES OF HAZOP

Different types of HAZOP studies are conducted depending on the problem's objective. HAZOP reports should follow standard procedures to make them valid under legal challenges (IEC, 2010; Crawley et al., 2000; Kyriakdis, 2003). The following list explains the types along with their applicability.

Process HAZOP: It is a technique developed to assess plants and process systems at their preliminary stage (test run condition). It is quite a common practice in the oil and gas industries.

Human HAZOP: It belongs to a group of specialized HAZOP studies. It is more focused on human errors than technical failures. They are usually conducted only on violations of work permits or reports of the bulk of near-miss events.

Procedure HAZOP: This is a review of procedures or operational sequences and is often classified as a Safe Operation Study (SAFOP). It is carried out while a significant deviation in the process line is proposed.

Software HAZOP: This is carried out to identify possible errors in software development, which controls the process flow or the production line of the plant. It is useful for CAD-CAM-controlled systems. It is helpful to analyze the hazards that may arise from the failure of automated control systems. It is critical and very useful to diagnose the electric and electronic control systems to avoid accidents arising from mechanical (or automated) faults. It is one of the common hazop studies carried out in oil and gas industries.

2.23   CASE STUDY: GROUP GATHERING STATION

Let us consider a case study of a Group Gathering Station (GGS). Location and intrinsic details of the GGS are masked for strategic reasons, but the study is carried out on a functional plant (Chandrasekaran, 2011c). The aim is to identify the hazards and operability problems of a GGS that have the potential to cause damage to the operation, plant, personnel, and environment. The main objective is to eliminate or reduce the probability and consequences of incidents in the installation and operation of GGS. PHA-Pro7 software is used to prepare the HAZOP worksheet in the present study. Figure 2.11 shows the PFD of the GGS considered for the study. The workings of the GGS are briefly explained in the following text to familiarize the reader with the process.

The well fluid emulsion, received at the limits of the GGS, is distributed into three production manifolds. From the Main Group Header, well fluid goes to the Bath Heat Treater for the first stage of separation of oil, gas, and water. Separated oil is subsequently stored in the Emulsion Receipt (ER) tanks, while the associated gas is separated and taken to the flare stack. Separated water is then drawn to the Effluent

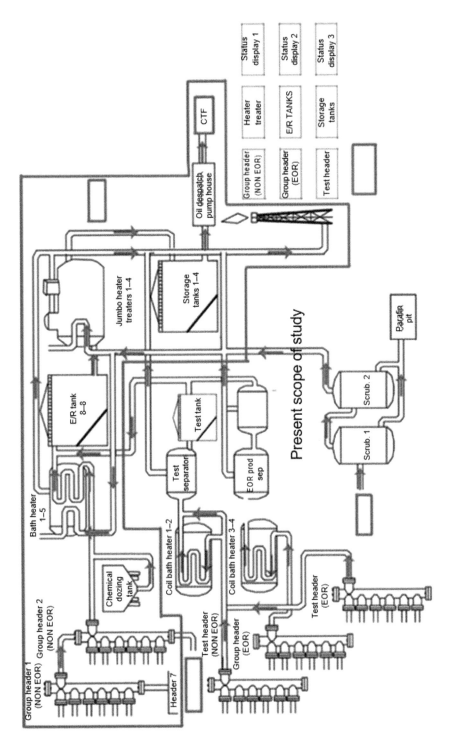

FIGURE 2.11 Group gathering station: Process flow diagram.

Treatment Plant (ETP) and disposed of after proper treatment. Oil from the ER tanks is fed to the Jumbo Heater Treaters through the Feed pumps for refinement. Oil and water are separated in the Jumbo Heater Treaters; separated oil is then pumped to the Common Tank Form (CTF). The flow of the process line is shown in Figure 2.11.

2.23.1 METHODOLOGY

1. A section of the plant (NODE) on the P&ID is identified.
2. **Design intent under normal operating conditions of the section is defined.**
3. Deviations from the design intent or operating conditions are identified by applying a system of predefined guidewords.
4. Possible causes, related consequences, and available safeguards shall be identified and reported.
5. Action(s) shall be recommended to reduce/eliminate the deviations; focus is kept on the consequences.
6. Discussions and actions shall be recorded in full detail.

3 Crude Oil and Natural Gas

Summary

This chapter discusses details about the origin of crude oil and natural gas with their geophysical properties. 2D and 3D survey maps, typically used in exploratory studies, are presented, explaining the terminologies useful for the readers. A brief discussion on non-conventional energy sources and green energy systems provides an overview and their relevance in the present context.

3.1 INTRODUCTION

Oil and gas are essential for ensuring global energy security, making a significant contribution to the world's energy needs (Agrawal et al., 2012). While oil and gas provide significant contributions to energy security, it pertains to the accessibility and dependability of energy resources at reasonable costs. They contribute significantly to the world's overall energy consumption. Diversification is crucial for ensuring energy security, as it allows for the development of a varied energy mix. Despite ongoing efforts to shift toward renewable energy resources, oil and gas still play a significant role in meeting the world's energy demands. This variety of energy sources helps the hazards associated with over-dependence on a single energy source. Transitioning to renewable energy sources is a global initiative aimed at combating climate change. However, throughout this transition, oil and gas can play a crucial role as an intermediary solution.

Oil and natural gas are created beneath the Earth's surface through the decomposition of prehistoric organisms by intense underground heat and microorganisms, a process that takes millions of years (Alvarado and Manrique, 2010). These resources are deposited in underground formations, where they tend to accumulate, giving rise to oil and gas fields. Deceased flora and fauna in terrestrial and marine environments descend into the depths of the sea and lake beds, along with microorganisms and sediment, where they amass over time (Bahadori, 2018). The accumulation of dead plants and animals gives rise to layers of sediment. Microorganisms underground decompose the accumulated deceased flora and fauna into organic compounds over millions of years, resulting in the generation of oil and natural gas. The term used to refer to these organic compounds is kerogen, while the rocks that contain a high concentration of kerogen are known as source rocks. Kerogen undergoes a conversion process to produce oil and gas resources in intense subterranean conditions characterized by elevated temperatures and pressures, which gradually ascend through underground fissures in rock layers. These resources migrate

DOI: 10.1201/9781003497660-3

69

vertically and accumulate in regions that meet the following two criteria: (a) below a compact layer with a dome-like shape known as cap rock, (b) in a porous rock formation known as a reservoir, the resources tend to accumulate due to the presence of void spaces.

3.2 ORIGIN OF CRUDE OIL AND NATURAL GAS

The investigations on the origin of oil have consistently focused on the analysis of organic matter and the identification of the processes involved in its transformation. Existing evidence unequivocally indicates that all detected accumulations of oil and gas have an organic origin. Chemofossils can be found in many substances such as oil, coal, oil shale, and bitumen (Baihly et al., 2010). These chemofossils are biomarkers that retain transitional bioorganic molecules' structure. Frequently, over 50% of the crude oil is comprised of biomarkers, which are not a blend but rather an essential component of the oils. The organic matter is primarily gathered in clay-rich marine sediments, often in a scattered distribution. The classification of organic matter is primarily divided into two categories: sapropelic and humic. The former was thought to have a significant impact on the generation of oil, while the breakdown of the latter led to the creation of coal, gas, and substances that can dissolve in water (thus, easily spread). Sapropelic matter undergoes decomposition, resulting in the formation of compounds that are in liquid and gaseous state, which include hydrocarbons. The thermal energy transfer and the accumulation of solar energy give rise to the decomposition process. The hydrocarbons, together with other compounds resulting from the decomposition of organic matter, and water are expelled from the shales into the reservoir formations. The hydrocarbons originating from the organic matter remain buoyant in the water medium due to gravitational forces and migrate until they become trapped within the reservoir. The marine origin of oil source rock seems evident; however, the reason for the first development of oil-bearing sequences in many countries being continental or near-shore marine Paleogenic and Neogenic rocks remains unknown. It was ultimately acknowledged that the only requirement was the existence of underwater sediments, which might be of continental or marine origin.

The geological history of the Earth has been divided into different units based on the events that occurred throughout each epoch. The geological timescale is primarily categorized into different eons, and each eon is subdivided into eras, and eras are further split into periods, epochs, and ages. The geologic timescale is divided into four eras, arranged in chronological order from oldest to youngest: Precambrian, Paleozoic, Mesozoic, and Cenozoic. An era is a segment of a geological time frame that is shorter than an eon while longer than a period. Radiometric dating establishes that the age of the Earth is around 4.5 billion years. The duration of time between this specific date and the emergence of bigger living forms in the fossil record is commonly known as Precambrian. Life existed in the Precambrian era. Around 600 million years ago, in the Paleozoic era, the emergence of life forms, including shellfish, began. Approximately 250 million years later, life was discovered in the onshore environment during the Carboniferous and Permian periods. The Mesozoic era comprises the Triassic, Jurassic, and Cretaceous timeframes. Around 225 million years ago, the emergence of dinosaurs signaled the commencement of the Mesozoic era. Mammals are distinctive of the most recent geological era, referred to as the Cenozoic, that commenced approximately 65 million years ago with the emergence of the first herbivorous and carnivorous mammals during a time referred to as the Tertiary. The Tertiary periods in the Cenozoic era consist of the Pliocene, Miocene, Oligocene, Eocene, and Paleocene.

Sedimentary basins are created over an extended period, spanning hundreds of millions of years, by the process of material erosion and the accumulation of organic matter and chemicals in aquatic environments. Over time, ongoing sedimentation takes place in the aquatic environment, resulting in increased weight and subsequent subsidence. The accumulation of organic matter and various components at distinct intervals spanning thousands of years will result in the formation of consistent layers of sediment in the basin. Tectonic activities or volcanoes induce the formation of faults. The elevated terrains' erosional activity, combined with further subsidence, ultimately results in the formation of a new low-lying land area. Subsequently, this area situated at a low elevation becomes inundated with water, so creating an additional aquatic habitat. Subsequently, further sedimentation occurs, resulting in an unconformity in the underlying layers of rock. An unconformity is a discontinuity in the geological records where there is a gap in the deposition of sedimentary layers, resulting in the separation of newer strata from older rocks. The layers lying above and below the unconformity may exhibit either parallel or non-parallel orientations. Ultimately, the displacement of land masses leads to the formation of folds and deformations. This phenomenon leads to the development of dense and intricate formations, ultimately resulting in the creation of a sedimentary basin. A stratum refers to a distinct layer of sediments, such as sand or shale, that is characterized by a clear alteration in lithology or a physical interruption in the sedimentation at its upper and lower boundaries. Strata is the plural form of stratum (Agrawal et al., 2012).

The Earth is comprised of three fundamental layers: crust, mantle, and core. Figure 3.1 shows the conceptual formation of the basin. The crust is the primary layer of utmost significance in petroleum geology. The Earth's crust consists of three fundamental rock types: igneous, metamorphic, and sedimentary. An igneous rock is a type of rock that forms from the solidification of molten material referred to as lava or magma. These rocks result from the process of magma cooling and solidifying. The cooling rate of this material significantly affects whether it exhibits coarse-grained or fine-grained structure. Differentiation can be made between volcanic rocks with larger grain sizes and volcanic rocks with smaller grain sizes. Volcanic rocks with a fine-grained texture are the result of magma cooling and solidifying quickly. On the other hand, coarse-grained rocks have a coarse texture because they take a much longer time to solidify.

Metamorphic rock is a type of rock that is created from existing rocks. This transformation occurs due to the effects of high temperature and pressure in the Earth's crust, resulting in changes to the mineral composition, chemical makeup, and texture of the rock. Slate is formed when shale undergoes metamorphism under low-temperature conditions. Under elevated temperature and pressure conditions, siltstone

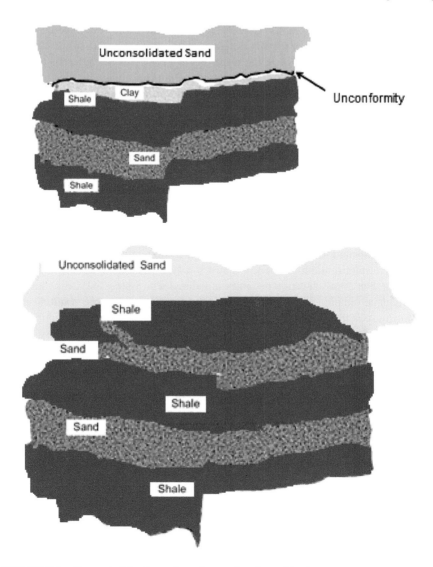

FIGURE 3.1   Concept of the formation of unconformity and basin.

and shale undergo complete recrystallization, resulting in the formation of schist and gneiss rocks. Slates, schist, gneiss, marble, and quartzite are different types of metamorphic rocks.

Sedimentary rock is created through the accumulation of either eroded fragments from older rocks found in upland regions or through chemical precipitation. The eroded materials are carried by water, wind, or ice to bodies of water such as the sea, lakes, or lowland areas. There, they settle and build up to create clastic sediments, which are fragmented in nature. Sediments undergo lithification through compaction, where the grains are compressed to form a denser mass than the original, and cementation, where mineral precipitation around the granular particles takes place post-deposition and combines the particles. Sediments undergo compaction and cementation following their burial beneath additional sediment layers. Sandstone is formed through the process of lithification, which involves the consolidation of sand particles. Limestone, on the other hand, is formed through the lithification of shells and other calcium carbonate-containing particles. The seawater's evaporation in confined areas can result in the accumulation of salt on the ocean floor. Evaporates are formed when anhydrites, calcium carbonate, and various salts are created in conditions where water is removed. Throughout this geological process, the entrapment of organic matter leads to the formation of the oil and gas that we are endeavoring to locate through the utilization of seismic techniques. Subsequently, these materials undergo a process of solidification, resulting in the formation of sedimentary rocks like conglomerates, sandstone, siltstone, and mudstone or shale. Limestone primarily consists of calcite and is primarily formed by calcium-carbonate-secreting organisms such as algae, corals, and animal shells, which are abundant in nature (Chong et al., 2010). Subsequently, chemical modification can transform calcite into dolomite. Important to the petroleum industry, sedimentary rocks are the primary type of rock where oil and gas accumulations predominantly occur. Sedimentary rock is exclusively the type of petroleum source rock.

Sedimentary basins are present worldwide, located at the peripheries of continental shelves. Hydrocarbons are derived from the source rock within a sedimentary basin. Hydrocarbons are produced through organic evolution. Oil and gas are formed through the continuous burial of significant amounts of organic matter (plants and animals) in deltaic, lake, and ocean environments. A "delta" refers to an accumulation of sediment that forms at the point where a stream or a river meets an ocean or lake. The organic debris is quickly buried in the sinking sedimentary basin. As sedimentation persists and overburden pressure intensifies from added weight, sediments carrying organic debris progressively descend further into the Earth. Typically, organic debris undergoes decomposition when exposed to oxygen. However, as the depth increases, the sediment acts as a shield, creating an environment devoid of oxygen, which helps preserve the organic matter. It facilitates the accumulation of organic matter instead of its decomposition by bacteria. The temperature within the Earth rises as depth increases. Consequently, sediments as well as the organic matter they hold experience an increase in temperature as they are progressively covered by younger sediments. Throughout millions of years, the accumulation of heat and pressure leads to chemical reactions that transform organic matter into kerogen.

As seen in Figure 3.2, Kerogen holds major geological significance, as it serves as the precursor for the formation of oil, gas, and coal. The formation of these products depends on the decomposition of specific organic debris as well as the pressure and temperature they have experienced over thousands of years. The rate of hydrocarbon generation is significantly reduced at temperatures below 150°F, while it reaches its peak within the temperature range of 225 to 350°F. With a continued rise in temperature, the source rock experiences increased heat, causing the separation of the chains of carbon and hydrogen atoms, ultimately resulting in the formation of heavy oil. With further increase in temperature, the heavy hydrocarbon undergoes conversion into lighter versions of light oil or gas. Gas can also be generated through the direct

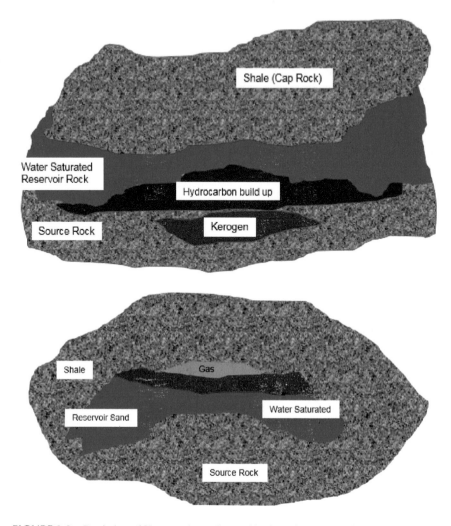

FIGURE 3.2   Depiction of Kerogen formation and hydrocarbon reservoir.

decomposition of kerogen derived from the lignocellulosic component of plants. The oil and gas generated through these procedures can exist in various combinations and are consistently blended with water. The hydrocarbon generated within the source rocks is subsequently transported to the pore spaces of rocks that are permeable. Due to their low density compared to water, these hydrocarbons migrate upwards through the permeable rock until they encounter a rock medium with no permeability (like shale), at which point they combine to form large quantities. An accumulation of hydrocarbon has now been created. The profitability of an oil and gas field can be determined by the reservoir size and the hydrocarbon accumulation within it. The interval of depths at which the generation of oil takes place is referred to as the "oil window," and it typically varies across different sedimentary basins. It is important to note that hydrocarbon deposits found in commercial quantities are categorized as pools, fields, or provinces.

Pool is the most basic form of commercially viable hydrocarbon accumulation in the underground. This hydrocarbon accumulation refers to a group of oil and gas located in the same subsurface reservoir sharing the same temperature and pressure conditions and confined within a single geological trap. A "hydrocarbon field" refers to a group of oil and gas pools that are connected through a common geological feature, either in terms of structure or stratigraphy. A province is formed by geological regions that contain a large number of oil and gas fields. A "reservoir" refers to an underground volume of rocks that are permeable and porous and can store fluids and facilitate their movement. At temperatures exceeding 500°F, the Kerogen undergoes carbonization, resulting in the cessation of hydrocarbon formation.

3.3 ESSENTIAL PROPERTIES: CHEMICAL AND RHEOLOGICAL CHARACTERISTICS

Understanding the characteristics of crude oil and natural gas is essential for developing and evaluating petroleum production systems. Oil properties comprise of gas-oil ratio (GOR), viscosity, density, compressibility, and formation volume factor. There exists an interconnection between viscosity, density, compressibility, and formation volume factor through GOR (Guo, 2007). Properties of gas encompass pseudo-critical temperature and pressure and specific gravity, which depend on composition and density, viscosity, formation volume factor, compressibility factor, and compressibility, which are contingent on pressure.

GOR refers to the quantity of gas (under standard conditions) that can dissolve in unit oil volume when the mixture is brought down to the reservoir at the current pressure and temperature. In general, the standard condition is specified as 60°F and 14.7 psi. At a specific reservoir temperature, GOR stays constant at pressures exceeding the bubble-point pressure (BPP). It decreases as the pressure drops below the BPP range. Solution GOR is determined in PVT laboratories. The GOR factor is commonly utilized in reservoir engineering for volumetric oil and gas calculations. It is also utilized as a fundamental factor in estimating various fluid properties, including oil density.

Viscosity describes a fluid's resistance to flow. Oil viscosity is a key factor in hydraulic and well-inflow calculations in the field of petroleum fluid production technology. The density of oil is the oil's mass per unit volume, measured in lb./ft^3. It is commonly used in hydraulic calculations, such as those for well-bore and pipeline performances. The oil density is dependent on pressure due to its gas content. API gravity is used to represent the oil density at standard conditions (stock tank oil). The compressibility of oil (or its isothermal compressibility, OIC) is defined as the fluid volume change following pressure change at an isothermal state. OIC under BPP is primarily influenced by the free gas compressibility due to their significant difference. A rise in the compressibility factor occurs at BPP occurs suddenly. The measurement of this property takes place in PVT labs. The property is primarily employed in reservoir simulation applications and to model well-inflow performances.

The formation volume factor (FVF) of oil refers to the ratio of oil volume in the reservoir at the existing temperature and pressure conditions to the oil volume in the stock tank, along with the dissolved gas. FVF is consistently greater than one due to the increased gas dissolution in reservoir conditions compared to stock tank conditions. At a specific reservoir temperature, FVF stays relatively constant at pressures higher than BPP. It decreases as pressure drops within the range of pressure below BPP. Oil FVF is determined in PVT laboratories. Oil FVF is commonly utilized in oil volumetric and well-inflow calculations. It is also utilized as a fundamental factor for predicting other fluid characteristics.

Gas's specific gravity is the ratio of its apparent molecular weight to that of air. Its critical properties can be calculated by applying the mixing rule based on the critical properties of the compounds present in the gas, similar to how the apparent molecular weight of the gas is determined. The critical properties of the gas evaluated in this manner are referred to as pseudo-critical properties. Viscosity is the degree of fluid's resistance to flow or deformation. Natural gas typically has low viscosity because it is in a gaseous state. Various factors like pressure, temperature, and composition influence the viscosity of natural gas. The compressibility factor of the gas is synonymous with the deviation factor. The value represents the extent to which the real gas differs from the ideal gas under specific temperature and pressure conditions. The density of a gas is the ratio of its occupied volume to mass and is influenced by parameters like composition, temperature, and pressure due to its compressible nature. The gas formation volume factor is the ratio of the volume of gas at reservoir conditions to the volume of gas at standard conditions. "Gas compressibility," or isothermal compressibility (GIC), refers to the change in fluid volume in response to a change in pressure at a constant temperature.

Similar compound classes are shared by crude oils and diluted bitumen; however, the proportions of these classes differ significantly. These variations are linked to significant disparities in chemical and physical characteristics. As per the analysis standards of the industry, the compounds are categorized into four primary classes: saturated and aromatic hydrocarbons, resins, and asphaltene. Light crude oils contain a high concentration of saturated hydrocarbons, making them less viscous and less dense. Crude oils that are denser and more viscous contain higher levels of other components, such as asphaltenes and resins, that consist of more polar compounds, often containing oxygen, sulfur, nitrogen, hydrogen, and carbon.

Even within medium or light crude oils, the proportions of certain compounds can differ greatly. The proportions of different compounds will vary based on the specific organic material in the sediments, the duration and intensity of heating of source rock, the presence of specific inorganic minerals that can catalyze reactions, the migration pathway details and distance, and the reservoir conditions. Under anaerobic conditions during the oil sands' formation, saturated hydrocarbons are highly biodegradable, aromatic hydrocarbons are less biodegradable, and resins and asphaltene are not biodegradable at all. Heavy crude oil, or bitumen extracted from oil sands, consists of the residue of a lengthy process in which microbial activity breaks down most of the saturates that can be metabolized. The fraction of saturated hydrocarbons in other crude oils is distinct from diluted bitumen due to the absence of molecules that can be easily metabolized. It is most prominently observed in chemical analyses that display the individual component distribution in crudes.

Aromatic hydrocarbons that have one or more aromatic rings are present in crude oils. Compounds possessing multiple rings are often known by the name polycyclic aromatic hydrocarbons (PAHs). The one-ring compounds are collectively known as BTEX, derived from the chemical names of benzene, toluene, ethyl benzene, and xylene, and are the most prevalent. Naphthalenes are the most prevalent two-ringed aromatic hydrocarbons. 3-ring phenanthrenes, dibenzothiophenes, and fluorenes, as well as the 4-ring chrysenes, belong to the other frequently measured groups. Naphthalenes and much bigger phenanthrenes have less volatility and solubility than BTEX. PAHs exist primarily as alkyl-substituted forms, with only a small portion being unsubstituted or parent forms.

Resins and asphaltenes found in diluted bitumen and heavy crude oils can separate from the oil and form black sludge, leading to issues such as blockages in well-bores, pipelines, and equipment. Additionally, the expense of refining rises in correlation with the quantities of asphaltenes and resins present. Medium and light crudes are preferred for those reasons. Broader use of heavy crudes has come into existence due to the added pressure on the supply and continuous advancements in the refining process.

Significant difficulties are faced by chemical analysts due to resins and asphaltenes. The variety of structures as well as the inclination of molecules to form larger aggregates make it challenging to ascertain basic properties such as molecular weight. Research has revealed that, in heavy residue, the number of carbon atoms present in the majority of individual molecules ranges from 30 to 70. They consist of a diverse assemblage of polycyclic molecules within that range. These molecules tend to adhere to each other, even at low concentrations, which is advantageous for asphalt as a paving material. The nano aggregates formed have masses that are twice to five times greater than the masses of the individual molecules they are made of. The observed molecular weights are higher than the actual molecular weights. Nitrogen, sulfur, and oxygen (referred to as heteroatoms) are more abundant in the fraction comprising asphaltenes and resins compared to the fraction comprising saturates, along with metals like iron, vanadium, and nickel. Therefore, the heteroatom content in bitumen is greater compared to other types of crude oils.

Raw bitumen's high density and viscosity necessitate heating or modification for transportation through transmission oil pipelines. To decrease the density and viscosity, a diluent needs to be mixed with bitumen to create a specialized blend with a viscosity under 350 cSt and a density below 0.94 g/cm^3. Industry-standard specifications for pipelines change seasonally based on the operating temperature. Diluents by themselves do not provide distinctive chemical or toxicological characteristics to diluted bitumen; instead, the more frequently utilized diluents are naturally existing combinations of light hydrocarbons. The light hydrocarbons are obtained from two sources: ultra-light crudes and gas condensates. Both have lower density and viscosity compared to synthetic crude oil, resulting in diluent-to-bitumen ratios of approximately 30:70. The specific combination of lighter hydrocarbons in the diluent will play a crucial role in spill response. If lighter compounds of the range C4-C8 are mainly present in the diluent, then they quickly evaporate in the case of a spill, leaving behind a thick and sticky residue.

Natural gas is a mixture of hydrocarbons composed mainly of light paraffins like methane (CH_4) and ethane (C_2H_6), which exist in gaseous form at normal atmospheric conditions. The mixture may also include additional hydrocarbons like hexane (C_6H_{14}), pentane (C_5H_{12}), butane (C_4H_{10}), and propane (C_3H_8). Heavier hydrocarbons in the reservoirs of natural gas are predominantly found in a gaseous state due to elevated pressures. They typically turn into liquid form when they reach the surface under atmospheric pressure and are extracted in the form of natural gas liquids (NGL), in gas processing plants or field separators. After being removed from the streams of gas, NGL can further be divided into different fractions, starting from the heavier condensates like hexanes, pentanes, and butanes, then moving on to liquified petroleum gas (LPG), which mainly consists of butane and propane, and finally, to ethane.

Additional gases often found alongside hydrocarbon gases include argon, helium, hydrogen, carbon dioxide, and nitrogen. Carbon dioxide and nitrogen are non-flammable gases that can be present in significant amounts. Nitrogen is chemically inactive, but in large quantities, it diminishes the mixture's heating value; hence, it needs to be eliminated to increase the heating value, decrease volume, and maintain consistent combustion properties. Natural gases frequently contain significant amounts of hydrogen sulfide (H_2S), and sometimes, other organic compounds of sulfur. Here, the gas is referred to as sour gas. Sulfur compounds are eliminated during processing due to their toxicity when inhaled, corrosive nature toward plant and pipeline infrastructure, and the pollution they cause when burned in products derived from sour gas. After removing sulfur, a small amount of pungent mercaptan odorant is consistently included in commercial natural gas to detect any potential leaks during transportation or usage. Natural gas extracted from a well has vapors of water along with it due to the presence of the formation of water in the reservoir. Its water vapor is partially condensed while being transported to the processing plant.

Comprehending the rheological characteristics aids engineers and operators in enhancing the design and functioning of systems used in the production, transportation, and processing of crude oil and natural gas. It also helps to tackle challenges such as flow assurance by preventing issues like wax deposition and hydrate formation that can disrupt operations. Accurate modeling of these properties is crucial for the efficient and safe handling of hydrocarbons across the entire supply chain. Crude oil viscosity is a crucial factor that impacts its flow characteristics. It is commonly quantified in centipoise (cP). The viscosity of crude oil varies greatly between different types of crude oils. Light crude oils typically exhibit lower viscosities, facilitating easier flow, whereas heavy crude oils tend to be significantly more viscous. The relationship between temperature and viscosity is crucial. Higher temperatures cause a decrease in crude oil viscosity, making it easier to transport. The viscosity of crude oil is greatly affected by fluctuations in temperature. Crude oils can experience increased viscosity and flow problems, like wax deposition, at low temperatures.

Crude oil density impacts its buoyancy, determining whether it will float or sink in water. Modifications in pressure and temperature can impact the density of crude oil. Certain heavy crude oils and bitumen may have yield stress, requiring a specific level of stress or force to start flowing. It is crucial to surpass this yield stress to pump and extract heavy crude oils. Crude oil demonstrates a steady flow behavior under constant conditions, with viscosity remaining relatively unchanged over time during continuous flow. Comprehending the steady flow characteristics is essential for pipeline design and efficient transportation. Crude oil can exhibit transient flow behavior during initiation, cessation, or abrupt alterations in flow circumstances. Changes in viscosity and flow properties can transiently change the efficiency of pumps and other machinery.

Thixotropy is the characteristic of decreasing viscosity over time when subjected to constant stress or shear rate. Certain types of crude oils may display thixotropic characteristics, impacting their flow behavior when being pumped or transported. Crude oils can exhibit viscoelastic properties, showing traits of both viscous fluids and elastic solids. Comprehending viscoelasticity is crucial for forecasting the behavior of crude oil under dynamic forces and deformations in extraction and processing. Several types of crude oils demonstrate shear-thinning characteristics, where their viscosity reduces as the shear rate rises. Its property is essential for pipeline transportation and pumping operations because oil becomes less resistant to flow at higher shear rates. When crude oil moves through pipelines or is subjected to pumping forces, its viscosity decreases, which facilitates its transportation. Its property is advantageous for effective transportation, particularly in high-flow situations. Shear thickening is rare in crude oils, but certain non-Newtonian fluids, particularly those with a higher number of solid particles or additives, can display shear thickening characteristics. As the shear rate increases, the viscosity of the crude oil also increases, resulting in greater flow resistance. Natural gas consists mainly of gaseous hydrocarbons, resulting in significantly lower viscosity than crude oil. The viscosity of natural gas can change depending on its composition, with heavier hydrocarbons leading to slightly increased viscosities. Under standard conditions, the density of natural gas is much lower than that of crude oil due to its gaseous state. Density fluctuates with changes in pressure and temperature, impacting its response to compression and expansion procedures. Natural gas is very compressible, allowing its volume to be greatly decreased when pressure is increased. Compressibility plays a crucial role in the conveyance of natural gas via pipelines and storage facilities.

Under conditions of high pressure and low temperature, natural gas can display non-Newtonian characteristics, departing from the typical correlation between shear stress and shear rate. Understanding non-Newtonian effects is crucial for accurately simulating natural gas flow under different conditions. Natural gas, being a gas, generally exhibits steady flow behavior under constant conditions. However, variations in pressure and temperature can impact its viscosity, influencing flow characteristics. Natural gas may experience transient flow behavior during changes in pressure and temperature, which can impact its flow dynamics. Understanding transient behavior is important for maintaining the integrity of pipelines and equipment. Under standard conditions, natural gas does not show a yield stress because it is in a gaseous state. Under severe conditions or when heavy hydrocarbons are present, flow resistance may occur. Thixotropic behavior is uncommon in natural gas, as it is typically associated with viscous liquids rather than gases.

Natural gas usually acts as a compressible fluid and does not show notable viscoelastic properties. Under high pressures and low temperatures, deviations from ideal gas behavior can occur. Shear thinning is uncommon in natural gas. Gaseous substances such as natural gas usually have low viscosity, which remains fairly constant regardless of the shear rate in normal conditions. Shear thickening is not a common occurrence in natural gas under standard conditions. Natural gas is a compressible fluid with consistently low viscosity, even when subjected to changing shear rates. While some rheological properties like steady flow behavior and transient flow behavior apply to both crude oil and natural gas, others like yield stress, thixotropy, and viscoelastic behavior may have more relevance to specific characteristics of crude oil. The rheological behavior varies based on the particular composition, temperature, and pressure of the crude oil or natural gas being considered.

Comprehending shear thinning and shear thickening behaviors is crucial for designing and operating systems used in the transportation and processing of crude oil and natural gas. These behaviors impact the choice of pumping equipment, the design and configuration of pipelines, and the overall effectiveness of the transportation process. Moreover, the rheological behaviors can be affected by the existence of additives or contaminants in crude oil.

3.4   OIL AND GAS RESERVOIR

Once hydrocarbons are formed, they move along pathways. Hydrocarbons are usually less dense than water and tend to move towards the surface. Petroleum will gather in sites where oil and gas are unable to move further to the surface. There are two main categories of traps: structural and stratigraphic. A structural trap occurs when the reservoir's geometry hinders the flow of fluids. Structural traps form when reservoir beds are deformed and potentially faulted into configurations that can hold economically valuable fluids such as oil and gas. Anticlines are a prevalent form of structural trap. Regional or tectonic activities can cause faulting and folding. Tectonic activities result from the movement of plates, whereas regional activities are demonstrated by the formation of salt domes. The creation of a structural trap through regional activity is referred to as a diapiric trap. Stratigraphic traps form when fluid movement is obstructed by variations in the characteristics of the rock formation. The formation undergoes changes that impede the upward movement of hydrocarbons. Stratigraphic traps can be formed due to thinning out of sand or porosity decline due to diagenetic alterations. Diagenesis is the alteration of the formation's lithology at lower temperatures and pressures in comparison with metamorphic rock formation. Diagenesis involves dolomitization, cementation, and compaction processes. "Dolomitization" refers to the chemical process where a magnesium atom replaces a calcium atom from calcium carbonate to create dolomite. The dolomite formed has a smaller volume than the initial calcite, leading to the creation of more porous space in carbonate reservoirs. Its type of porosity is referred to as secondary porosity. Combination traps are formed by a mix of stratigraphic as well as structural features.

Petroleum reservoirs are typically located in sedimentary rocks. Petroleum is occasionally discovered in rock formations of igneous or metamorphic type, which are fractured in nature (Gao and Du, 2012). These rocks form under intense heat and pressure, which are not conducive to the creation of petroleum reservoirs. Typically, they lack the interconnected pore space and permeability required for hydrocarbons to move toward the well-bore. A metamorphic rock, although it may have started as sandstone, has undergone heat and pressure. Any hydrocarbon fluid that may have previously filled the pores is evaporated. The most probable type of rock for holding significant amounts of hydrocarbon is sedimentary formation. For a petroleum reservoir to form, several essential components are required. To begin with, a source rock containing hydrocarbons must be available. Hydrocarbons are believed to originate from remnants of marine organisms. Remains gather in a sedimentary setting like shale, which transforms into a source rock. The source rock's pressure and temperature must be appropriate for oil or gas generation from the organic mixture. Suboptimal conditions can hinder the generation of hydrocarbons. At too-high temperatures, overcooking of the organic matter, which is decaying, happens. It could lead to the creation of gas as well as carbon residue. Another important factor is the existence of a reservoir formation as well as a pathway connecting it to the source rock. Rock is classified as reservoir rock if it contains fluids within its volume and sustains economically viable flow rates. Hydrocarbons are usually created in formations that are not easily accessible using current production methods. Hydrocarbons must possess the ability to move into the well-bores to be producible. The rate of flow should be sufficient to ensure the economic viability of the wells. Porosity and permeability are two crucial factors that determine the economic feasibility of the reservoir. Porosity is the proportion of space volume to the overall volume. Porosity is a crucial factor in determining the capacity of a reservoir to store fluid.

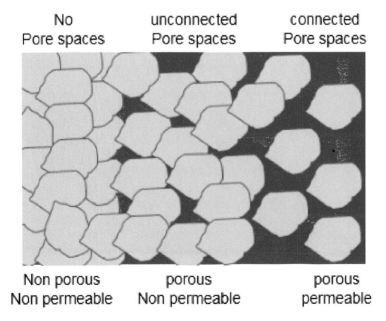

FIGURE 3.3 Depiction of porosity and permeability.

Permeability is the degree of interconnectedness of the pore spaces which enables the movement of fluids. Figure 3.3 shows a graphical representation of porosity and permeability.

Unrestricted hydrocarbon fluid will naturally ascend due to buoyant forces and other driving mechanisms if not contained. Fluid confinement requires two factors: vertical restriction by a cap rock that has no permeability and lateral containment by a trapping mechanism. Figure 3.4 shows the anticline and structural trap along the plane of fault while Figure 3.5 shows the stratigraphic trap. Timing supersedes all these factors. Without proper timing coordination, no action can take place. If a trap

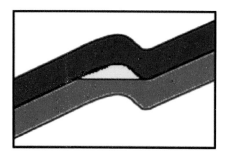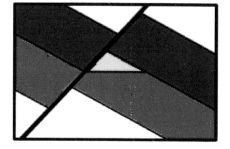

FIGURE 3.4   Structural trap: Anticline and along the plane of fault.

FIGURE 3.5   Stratigraphic trap.

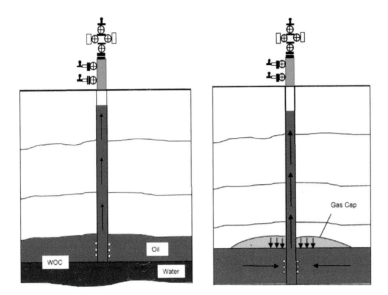

FIGURE 3.6   Water drive and gas-cap drive.

forms a million years after oil has passed through a reservoir, not much oil will be found except for possibly oil-stained rock, even if the source rock offers oil in billions of barrels.

Hydrocarbon accumulations are categorized as oil, gas-condensate, or gas reservoirs based on the original reservoir condition. An oil exceeding its bubble-point limit of pressure is referred to as under-saturated oil as it can dissolve additional gas at the specified temperature. An oil that has reached its BPP is referred to as a saturated oil, as it is unable to dissolve any more gas at the specified temperature. Undersaturated oil reservoirs exhibit single-phase flow with liquid, while saturated oil reservoirs have two-phase flow with liquid oil and free gas. Wells belonging to a single reservoir may be classified as oil, condensate, or gas wells based on the GOR they produce. Gas wells have a GOR value greater than 100000 scf/stb. Condensate wells have a GOR between 5000 and 100000 scf/stb while oil wells have a GOR below 5000 scf/stb. Oil reservoirs may be categorized based on the boundary type, which influences the driving mechanism. Figure 3.6 shows the water drive and gas trap drive, while Figure 3.7 shows the dissolved gas drive.

A dissolved-gas-drive reservoir is also known as a solution-gas-drive reservoir and volumetric reservoir. The oil reservoir contains a constant volume of oil and is enclosed by boundaries that do not permit any flow as in the case of faults or pinchouts. The drive mechanism is established on the fact that the reservoir gas is maintained in solution within the liquid (oil and water). The gas in the reservoir exists in a dissolved form in a solution with oil and water from the reservoir under atmospheric conditions. As compared to water-drive and gas-drive reservoirs, the expansion of dissolved gas in oil serves as a less effective driving force in the volumetric reservoir. Gas escapes from the oil in regions where the oil pressure falls beneath BPP,

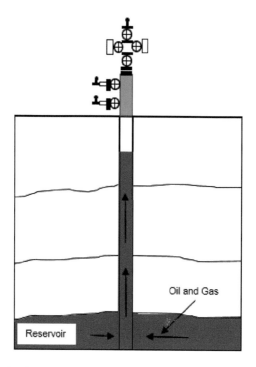

FIGURE 3.7   Dissolved-gas drive.

resulting in an oil-gas two-phase flow. Pressure maintenance at an initial stage is typically favored to enhance oil recovery in these types of reservoirs (Sheng, 2013).

A gas-cap-drive reservoir utilizes a drive mechanism where gas separates from the liquid in the reservoir and accumulates at the top to create a gas cap. Therefore, the oil beneath the gas cap can be extracted. Removing gas from the gas cap at the beginning of production will cause a rapid decrease in reservoir pressure (RP). Occasionally, an oil reservoir experiences both water- as well as gas-cap-drive. In the case of a water-drive reservoir, the oil-zone is directly linked to the aquifer forming a continuous pathway to the surface groundwater system. The pressure exerted by the water column pushes the hydrocarbon fluids to the reservoir top, where they are contained by the impermeable trap boundary. This pressure will push the hydrocarbon fluids toward the well-bore. An active water drive will help maintain RP for a longer period compared to other drive mechanisms with matching oil production. Water-drive reservoirs are classified as edge-water-drive and bottom-water-drive reservoirs. The former is the preferred type of reservoir over the latter. The RP can be maintained at its initial position over the BPP to ensure single-phase liquid flow in the reservoir, maximizing well productivity. In an edge-water-drive reservoir, a steady-state flow regime would persist for an extended period before water breakthrough occurs in the well. Bottom-water-drive reservoirs are less favored due to water-coning issues impacting oil production economics through water treatment and disposal challenges.

3.5 SEISMIC SURVEYS FOR OIL AND GAS EXPLORATION FIELDS

Seismic surveys play a crucial role in the oil and gas industry. They are a valuable tool for industry professionals to discover viable areas for exploration, evaluate the amount of resources present, minimize uncertainty, and accurately measure reserves. In essence, they enable educated decision-making that leads to successful exploration endeavors. Seismic surveys have emerged as the predominant method employed by several exploration companies around the globe, encompassing both onshore and offshore operations. They have significantly reduced the expenses associated with discovering oil and gas reserves while also enabling exploration in areas that were previously inaccessible using alternative methods. It has brought about a revolutionary transformation in the sector.

A seismic survey is performed by generating a shock wave, known as a seismic wave, on the ground surface along a pre-set path using a source of energy. The seismic wave propagates through the Earth, undergoes reflection by subterranean formations, and subsequently, returns to the surface, where it is detected by receivers. Seismic waves are generated through the use of either small explosive charges detonated in shallow holes, known as shot holes, or by huge trucks equipped with heavy plates that induce vibrations on the ground. A geophysicist can use the analysis of seismic wave reflection times to create maps of subsurface formations and anomalies. It allows them to make predictions about areas where there may be substantial quantities of oil and gas trapped, which is useful for planning exploratory activities.

The process of seismic acquisition begins with a well-defined objective of specific features within the subsurface that geoscientists aim to capture in their imaging. The primary goal of seismic data acquisition is to procure data that can be correlated with the structural representation of the underlying geology to identify oil and gas reservoirs (Cipolla, 2009; Cipolla et al., 2011). Before conducting seismic acquisition, a surveyor examines and maps the specific area where exploration activities will occur and precisely designates the seismic line. The lines function as conduits for the preparation of shot holes and the installation of cables for recording purposes. The lines are established using Global Positioning System (GPS) and total stations. The survey lines typically create grids with a spacing of 500-500 meters.

The "receiver line" refers to the line where the groups of geophones are positioned at consistent spatial intervals. The receiver line is alternatively referred to as in-line. The "receiver interval" refers to the spatial separation between two receiver groups situated along a common receiver line. The "receiver-line" interval refers to the distance separating two successive receiver lines. "Receiver density" refers to the quantity of receivers per unit area. The receiver density is determined by the number of receiver lines and the number of receivers in a kilometer.

The act of setting off an explosive, specifically dynamite, is commonly known as a seismic shot. Shot points are the specific surface locations on the Earth where a seismic shot is initiated. The "shot-point interval" refers to the distance separating consecutive shot-points along a single receiver line. The "shot-line interval" refers to the distance between two consecutive shot lines. The shot line is alternatively referred to as cross-line. The "source density" refers to the quantity of shots distributed over a given surface area. The shot density is calculated from the number of source lines in a kilometer and the number of sources in a kilometer.

"Swath" refers to the division of the area of seismic survey into six receiver lines. "Salvo" refers to the total number of shots fired in a single shot line. If each shot is 40 meters apart and there are 100 shots fired in a shot-line, then the total length of the shot-line would be 100 multiplied by 40, resulting in 4000 meters, or 4 kilometers. "Azimuth" is the precise term used to describe the direction of the line connecting the shot and the receiver. The template is a representation of all currently active receivers associated with a specific shot point. It is important to note that the shot point may be located either within or outside the template. Patch is a method of acquiring data where the source lines and the receiver lines are not parallel to each other.

"Receiver-line move-up" is a term used to describe the process of shifting the template to the subsequent shot-line after finishing the shots per salvo. Shot-line move-up is the phenomenon occurring while the template touches the boundary of the survey area and transitions across the survey to initiate a new receiver-line move-up. Figure 3.8 shows different seismic acquisition parameters that are vital for the survey.

A seismic source is a device capable of emitting acoustic energy to collect seismic data. Seismic sources include explosive charges, vibrators, and airguns. Explosive energy sources, such as dynamite, are utilized to acquire onshore seismic data. Dynamite is employed as the seismic source to produce acoustic waves that are transmitted into the sub-surface. Typically, the seismic shots are positioned beneath the weathered layer of the Earth, which has a lower velocity. It enhances the connection between the seismic source and the sub-surface while preventing issues caused by the highly fluctuating seismic velocities in the weathering layer. Drilling of shot holes is performed at specific locations called shot points, into which the explosive charges are then inserted. For creating shallow pattern holes, typically with a depth of 3.5 to 4 meters, drill casing and hand auger are employed. The shot holes in swamp terrain typically have a depth ranging from 45 to 60 meters. The 2 kilograms of explosives are then inserted into the drilled shot holes. Every shot position is assigned a number, and the precise location of each shot point is mapped accurately. Shots from air-shooting and vibrators, other alternative seismic sources, involve shots that take place on the Earth's surface. The quantity and placement of shots are meticulously planned to enhance the downward propagation of energy and diminish energy propagation in other directions.

It is important to consider that the size of the explosives utilized is contingent upon the specific goal of the survey; specifically, the target reflectors' depth. The seismic source with a mass of 1 kilogram releases approximately 5 megajoules of energy. Certain oil companies employ 2 kilograms of dynamite as their seismic energy source. Therefore, the amount of released energy per shot is dependent on the size of the explosive, shot hole depth, and the conditions of the local ground. The precise coordinates of each shot position are meticulously documented, along with the measured depth of the shot-hole. The acquisition of seismic data necessitates the use of these parameters for accurate processing. After marking the shot positions, geophone groups are placed at regular intervals, usually extending over a distance of 3-6 kilometers, either around or to a single side of the shot position. These geophone

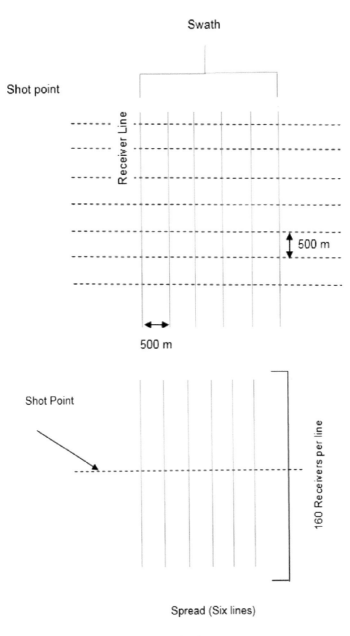

FIGURE 3.8   Seismic acquisition parameters for the survey.

groups are then connected by wire to the recording truck, and the acquisition job commences. As the shot position progresses along the line, various segments of the geophone groups are activated by the recording instruments. It ensures a consistent range of offsets for each shot. The "offset" refers to the precise distance between the shot and the receiver. Eventually, the receiver group needs to be relocated to preserve the same offset range in the "live" section. It is important to observe that variations in the velocity and density of sound within a specific rock layer result in the refraction and reflection of seismic waves generated by a seismic source. The variation of these characteristics at any interface between two distinct types of rocks results in the reflection of a portion of the seismic energy back toward the surface. The seismic section is generated by recording these reflections over time.

Seismic recording involves the use of a device capable of detecting acoustic energy, which can be in the form of ground motion or a pressure wave in fluid. This energy is then converted into an electrical impulse. Geophones, which are basic electromagnetic devices, are used to record onshore seismic data. A geophone is employed during onshore seismic acquisition, and offshore seismic acquisition is deployed on the seabed. The geophone detects the velocity of the ground caused by acoustic waves and converts this motion into electrical signals. It produces an electrical current that is directly proportional to the particle velocity of the Earth. It exits the wires linked to the geophones and is directed toward an analog-to-digital converter to generate the digital data. It can be transmitted through a cable to the recording unit and converted into digital format there. Alternatively, it can be obtained from the receiver groups as analog data and converted into digital format at that location before being transmitted via a cable to the recording unit.

It is important to acknowledge that geophones are capable of detecting motion alone in the vertical direction. The energy detected by a single geophone is minimal. Hence, a cluster of many geophones is arranged in an array encircling the position of the central receiver. This arrangement is referred to as a channel. The geophone within an array is subjected to analog summation for generating the output signal that is then directed into the central receiver. It signifies the connection of a single channel to the recording truck. During 2D seismic acquisition, it is typical to simultaneously record channels of the order of 120-240, while in the case of 3D seismic acquisition, it is typical to simultaneously record channels of the order of 500-2000. It is important to understand that arranging geophones in an array enhances the overall signal output of the group and also optimizes the geophones to amplify energy from below while reducing energy from the sides, such as ground roll.

Seismic traces are the recorded data obtained from a single shot point at a certain receiver location. Seismic traces are documented based on the progression of time. The term "two-way travel time" accurately refers to the duration that acoustic waves consume to propagate into the subsurface and reach back to the surface. The unit of measurement is either seconds or milliseconds. The aggregated set of traces obtained from a single shot point is often documented as a unified entity known as a seismic field record. The field record's horizontal scale is represented by the receiver number, which can be converted into feet/miles or meters/kilometers. The vertical scale represents the duration of travel in both directions. A field record typically comprises 240 seismic traces collected at 25-meter intervals on the surface, resulting in a substantial amount of data being captured during a seismic survey. The arrangement of several traces in their accurate spatial placements results in the creation of the ultimate "seismic section," which offers geologists and geophysicists a structural representation of the subsurface. In the field of 3D seismic processing, the survey region is partitioned into a grid with dimensions of 25 by 25 meters, which

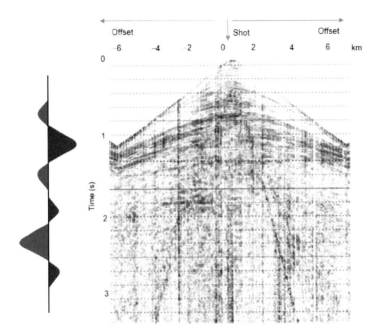

FIGURE 3.9 Typical seismic section.

is referred to as a bin. A bin is a geometric region, typically square or rectangular, that encompasses all data points associated with a specific common midpoint (CMP). Geophysicists consider seismic traces to exist precisely at the halfway between the source and receiver distances. The fold of coverage in a 3D survey refers to the quantity of traces included within a bin that will be combined through stacking. The fold values commonly observed in contemporary seismic data vary between 60 and 240 for 2D seismic data, while for 3D seismic data, the values vary from 10 and 120. Figure 3.9 indicates a typical seismic section, while Figure 3.10 shows a typical short line, receiver line and bin.

The "common mid-point (CMP) fold" refers to the quantity of traces that are linked to a specific mid-point location and are combined during the stacking process to generate a single trace. Fold-map, also referred to as multiplicity-map, displays the aggregate extent of coverage for a 3D seismic survey. The "nominal fold," also known as the full fold, of a 3D survey, refers to the fold achieved at the maximum offset. As a result of the reflection method, the fold is not optimal at the edge of the 3D survey.

3.5.1 Fold of Coverage Calculations

Receiver-line fold (R_F) is calculated using the number of recording channels per receiver line, represented by η_{RC}, the spacing or interval between the receivers, denoted by ζ_{RS}, and the shot-line interval, referred to as ζ_{SS}.

$$R_F = \eta_{RC} * (\zeta_{RS}/2) * \zeta_{SS} \qquad (3.1)$$

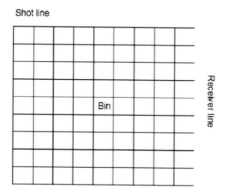

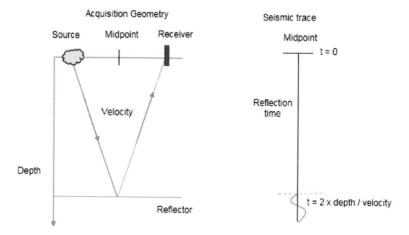

FIGURE 3.10   Typical short-line, receiver line, and bin.

Shot-line fold (S_F) is calculated using the number of shot points in a shot line or the number of shots in a single salvo, represented by η_{SC}, the spacing or interval between the shot points, denoted by ζ_{SS}, and the receiver-line-interval or receiver-line-spacing, referred to as ζ_{RS}.

$$S_F = \eta_{SC} * (\zeta_{SS}/2) * \zeta_{SS} \tag{3.2}$$

The total fold (T_F) is calculated from the receiver-line fold (R_F) and shot-line fold (S_F) and the total multiplicity (T_M) is the representation of the total fold in percentage.

$$T_F = R_F * S_F \tag{3.3}$$

$$T_M = T_F * 100\% \tag{3.4}$$

Offshore seismic surveys are performed by employing specialized vessels that pull one or more cables containing a sequence of hydrophones at consistent intervals.

Marine seismic surveys may employ a trio of seismic vessels. One of these vessels serves as the primary vessel for shooting, recording, and processing. To enhance coverage, the remaining two vessels towed extra streamers. These two additional vessels capture seismic signals; however, they are responsible neither for shooting nor for data processing. The streamers' locations are precisely determined using the GPS. It is important to acknowledge that a seismic vessel can acquire data at a rate that is tenfold higher than the most proficient onshore seismic crew. It is because the hydrophone arrays are pulled behind the boat, so obviating the necessity of constantly deploying new spreads. Airguns serve as seismic energy sources in the process of collecting offshore seismic data. This device expels a powerful burst of densely pressurized air into the adjacent water. Airguns are sometimes utilized onshore as a power source in water-logged pits to acquire vertical seismic profiles. After the initial release of energy, the pressure interaction between the air bubble and the water results in the bubble's oscillation while it moves toward the surface. The difference in amplitude and time between the bubble oscillations is contingent upon the gun's depth and the dimensions of its primary chamber. The gun's depth is estimated from the intended bandwidth while the oscillations of the gas bubble, created by the airgun, produce consecutive pulses resulting in source-generated noise. An individual airgun is not an impeccable source. Utilizing an assortment of airguns with varying sizes of chambers and shooting them concurrently enhances the efficiency and elevates its status as a superior source. Strategic utilization of several airguns can induce harmful interference of bubble oscillations and mitigate the bubble noise. A cage, also known as a steel enclosure, can be utilized to effectively disperse energy and minimize the occurrence of the bubble effect when surrounding a seismic source. Each airgun array consists of three airgun strings, depending on the offshore acquisition design. Each string consists of nine airguns. The airgun array has an average length of 23 meters. A streamer is utilized for the acquisition of marine seismic data.

A 3D seismic vessel can deploy between 1-12 streamers and firing arrays. These devices offer the necessary high data densities for marine seismic acquisition. The streamers are positioned slightly below the water's surface and are positioned at a specific distance from the vessel. The hydrophone, which is used for recording, is housed within the streamer that is pulled after the seismic vessel. The hydrophone monitors variations in pressure resulting from the reflection of sound waves off the geological layers beneath the sea bottom, caused by the airguns. Hydrophones employed in offshore recording utilize a pressure-sensitive apparatus to capture the incoming energy. Hydrophones are interconnected in clusters within the streamer and can be positioned at intervals of 6 meters. The intricate electronics inside the streamer process the signals coming in from a collection of hydrophones, and subsequently, transform the resulting voltage into a digital representation. The values for all groups are multiplexed, meaning that their corresponding numbers are interleaved, and then, transmitted digitally through a limited number of cables to the recording devices. The final digital recordings are ultimately transmitted to the processing center, which may also be located on the vessel. The streamer's bird and other positioning devices are utilized to provide real-time data on the precise location of all the hydrophone groups of all the streamers during all the shoots. A tail buoy, which is a buoyant item, is employed to indicate the streamer's end. The front end of the streamer is linked to the seismic vessel using a configuration of buoys and

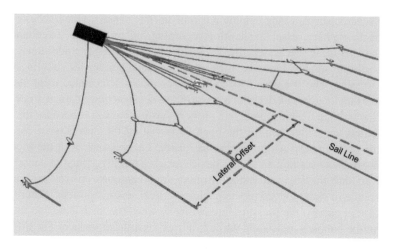

FIGURE 3.11   Aerial view of marine seismic data acquisition.

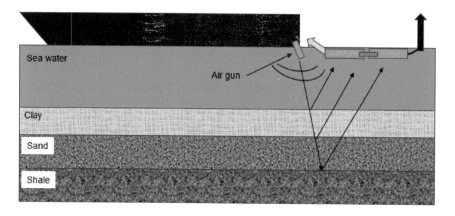

FIGURE 3.12   Side view of marine seismic data acquisition.

flexible parts with elastic properties. It effectively eliminates any extraneous sound that may arrive from the seismic vessel at the streamer. The tail buoys enable the seismic data acquisition team to track the position and orientation of the streamers. It is affixed to the cable's end and may include a GPS receiver and radar reflector of its own to accurately determine its location. Figure 3.11 shows an aerial view of marine seismic data acquisition, while Figure 3.12 shows the side view.

In a 2D seismic survey, as seen in Figure 3.13, both the shot and receivers are aligned along the same line. The subsequent line is located many kilometers apart. In simple terms, 2D seismic data is collected one at a time, while 3D seismic data is obtained by acquiring numerous closely spaced lines simultaneously. Figure 3.14 shows a typical 3D seismic survey. A 2D seismic data provides limited coverage of the sub-surface. Off-line reflections are accompanied by issues. These reflections are often referred to as sideswipe and are identical to the reflections exactly underneath.

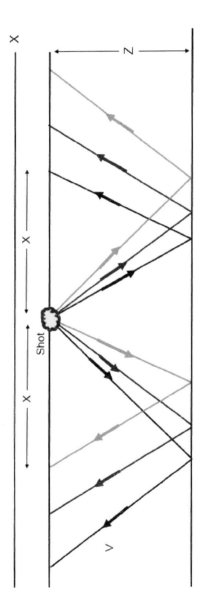

FIGURE 3.13 Two-dimensional seismic survey.

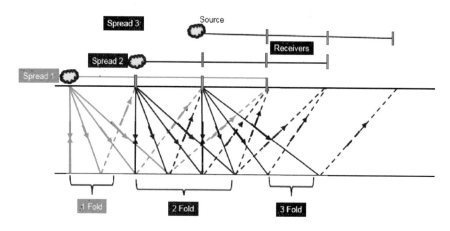

FIGURE 3.14   Three-dimensional seismic survey.

Sideswipe refers to an occurrence in 2D seismic data where a geological feature, like a fault or anticline/syncline structure, to name a few, which is not aligned with the seismic section, becomes visible. Figure 3.15 shows 3D baseline and monitor surveys. An appropriately migrated 3D survey shall be free from any instances of sideswipes. To address both issues of coverage and sideswiping, it is necessary to encompass the entirety of the survey region characterized by a collection of closely positioned seismic lines, resulting in the creation of a three-dimensional seismic data volume. In a 3D survey, each shot is positioned with a grid consisting of numerous lines of receivers. The next line of receivers is spaced several tens of meters distant. Put simply, a three-dimensional seismic volume is generated by capturing a dense grid of two-dimensional lines and then filling in the gaps between the lines to form a three-dimensional data volume, sometimes called a cube. Both onshore and off-shore 3D data are obtained using various arrays of sources and receivers to enable the acquisition of huge data volumes. The 3D seismic approach frequently enhances the density of data. It addresses numerous issues encountered in 2D sections, like reflections occurring outside the plane or sideswipe incidents. 3D seismic data offers comprehensive insights into the distribution of faults and the underlying structure of the Earth's subsurface, which is not possible with 2D data. This method provides a three-dimensional representation of the Earth, enabling versatile analysis of the data. The outcomes entail an enhanced comprehension of the composition and characteristics of the Earth's subsurface, hence increasing the likelihood of discovering economically viable deposits of hydrocarbons.

A 4D seismic survey is a method of collecting 3D seismic data at various intervals to analyze the evolution of hydrocarbon reserves over time. Alterations can be detected in dynamics and saturation of fluid as well as in pressure and temperature. The oil and gas industry uses 3D-time-lapse seismic surveys to observe the movement of fluids within a reservoir throughout production. It is done by performing an initial seismic survey and then conducting subsequent during the lifespan of the reservoir. When such surveys are conducted repeatedly in this manner, they are commonly known as 4D seismic surveys. Usually, the processing of 4D seismic

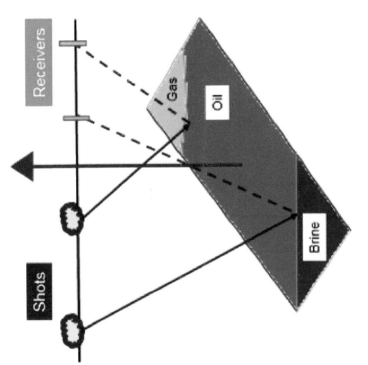

FIGURE 3.15 Three-dimensional baseline and monitor survey.

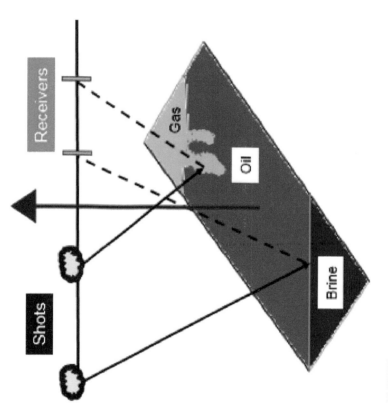

FIGURE 3.15 (Continued)

data involves subtracting the data obtained from the initial 3D survey from the data obtained during the subsequent 3D survey. The reservoir's change is determined by the disparity between the two. If there has been no alteration throughout the specified time frame, the outcome will be the value of zero. The utilization of a 4D survey will enable geoscientists and reservoir engineering experts to enhance the accuracy and efficiency of static and dynamic models of intricate reservoirs by employing cutting-edge computer technologies. The outcome of doing such a survey will be an improved extraction of hydrocarbon from the field, more optimal placement of injection and production wells, and decreased expenses associated with drilling, leading to an extended lifespan of the field. A 4D survey offers data that facilitates the effective administration of hydrocarbon fields. For instance, one can monitor the growth of gas-cap to effectively control production and prevent it from reaching the producing wells, hence minimizing the decrease in oil production rate. Additionally, it serves as a reference point for the progress of additional oil and gas resources that have been located nearby. Utilizing a 4D survey can effectively identify previously undiscovered reserves of hydrocarbons within a reservoir.

3.6 NON-CONVENTIONAL SOURCES OF OIL AND GAS

Evaluating conventional and unconventional resources is crucial for utilizing scientific and technological knowledge from the domains of geology, geophysics, petrophysics, and engineering to efficiently extract hydrocarbons from beneath the surface. While conventional resources like heavy oil and oil sands have been utilized for a while, the extensive extraction of hydrocarbons from unconventional sources existing deep down the Earth like shale gas and shale oil is a more recent development (Manchanda and Sharma, 2014). Conventional hydrocarbon resources usually gather in advantageous structural or stratigraphic traps where rock formations have commendable porosity and permeability but are enclosed by an impermeable layer that stops hydrocarbons from leaking out (Wang et al., 2015). The advantageous subsurface structures contain pathways that connect the reservoirs with the source rock formations for the migration of hydrocarbon fluids. The formations have high-quality reservoirs and typically do not need extensive stimulation to extract hydrocarbons. Unconventional resources have lower reservoir quality and are more challenging to extract hydrocarbons from as they are located in tight rock formations (Mayerhofer et al., 2010). Conversely, these resources are plentiful in the Earth. Unconventional resources encompass diverse geological formations such as oil and gas shales, heavy oil and tight gas sands, coalbed methane (CBM), and gas hydrates. Non-conventional hydrocarbon resources are those that do not fall under the category of conventional resources (Sivakumar et al., 2020). For instance, tight basement formations with lower permeability containing oil or gas are considered unconventional resources. Categorizing a reservoir into a particular class is often challenging. CBM, gas hydrates, and oil and gas shale are classified as unconventional resources, while tight gas sandstones are considered conventional (Sondergeld et al., 2010; Kok et al., 2010; Kraemer et al., 2014).

Tight gas sandstone reservoirs, along with other tight formations, exhibit similarities to conventional reservoirs (Passey et al., 2010; Pilisi et al., 2010). They can be classified as conventional unconventional resources or non-source rock unconventional reservoirs. Source rock unconventional reservoirs include shale, which are also referred to as deep unconventional reservoirs (Rabia, 1985). Uncertainty can also exist in a particular reservoir, where a shale formation might hold both oil and gas or a hydrocarbon-rich geological formation may consist of different lithofacies, including shale as well as other lithologies having coarser grains. Permeability levels ranging from low to ultra-low and porosity levels ranging from low to moderate are key features of unconventional reservoirs. Therefore, extracting hydrocarbons from these reservoirs necessitates the use of extraction technologies distinct from those used for conventional reservoirs. To recover significant amounts of hydrocarbons from an unconventional reservoir, it is necessary to stimulate it to achieve a satisfactory flow rate. Unconventional reservoirs typically have permeability levels lower than 0.1 mD, while reservoirs with permeability exceeding 0.1 mD are commonly classified as conventional. The permeability of a reservoir is not constant and is typically highly heterogeneous in both unconventional as well as conventional reservoirs, making it challenging to determine definitively. Due to their low permeability, shales are usually seen as source rocks or seals in most of the conventional reservoir settings. However, they can also serve as significant reservoirs.

Shales are a type of sedimentary rock that are abundantly found on Earth; however, not all shales contain hydrocarbons, and not all shale reservoirs are of the same quality. No single geological or petrophysical parameter can definitively indicate high oil and gas production. Instead, two crucial categories of variables play a significant role: the quality of the reservoir and the quality of the completion (Miller et al., 2011; Cipolla et al., 2011). Reservoir quality refers to the potential for hydrocarbons, the volume of hydrocarbons present, and the rock formation's ability to deliver hydrocarbons. Completion quality refers to the capacity to generate and sustain fracture surface area, and thereby, is an indication of stimulation potential. Key factors affecting reservoir quality are total organic content (TOC), fluid saturations, thermal maturity, organic matter content, mineral composition, lithology, formation pressure, effective porosity, and permeability (Passey et al., 2010; Onajite, 2013). Completion quality relies significantly on the geomechanical properties and mineral composition of the formation, such as the occurrence and features of the natural fractures, the ability of the rock formation to be fractured, and in-situ stress regimes. Because the unconventional rock formations are tight, the process of developing these resources differs significantly from that of conventional resources. Extracting hydrocarbons from tight formations typically involves drilling numerous wells and utilizing advanced techniques such as horizontal drilling as well as hydrofracking. Horizontal drilling combined with multistage hydrofracking has been essential for the cost-effective extraction of hydrocarbons from various shale and tight reservoirs.

Key variables and parameters in these new technologies are patterns of horizontal wells, design of hydrofracking, count of stages in hydrofracking, and perforation clusters. Multiple well-spacing pilots' hydrofracking operation schemes are often utilized to optimize well placement (Waters et al., 2009; Warpinski et al., 2014). Uncertainty associated with the development of unconventional resources is elevated due to the intricate subsurface formations. The evaluation and development of unconventional resources necessitates a comprehensive multidisciplinary method.

An interdisciplinary process can effectively combine geological, petrophysical, geophysical, and geomechanical data to enhance reservoir property characterization, prioritize key parameters, improve production efficiency, and address uncertainty. An integrated workflow can effectively capture the key features of unconventional resources and provide a quantitative method to develop these fields effectively. Studying various factors such as TOC, mineralogy, lithology, pore structure, anisotropy, natural fractures, rock mechanics, in-situ stresses, impact of structures and faults, and reservoir production interaction is necessary. To fully characterize these formations, it is necessary to combine the data of different types such as core, borehole, well-log, seismic, and engineering. Exploration and drilling in such fields lack a comprehensive approach, leading to subjective decision-making and high levels of uncertainty and risk. Three methods to decrease uncertainty in field development are obtaining sufficient data, employing advanced technologies, utilizing scientifically valid inference for analyzing diverse data, and combining various disciplines in geoscience (Ma et al., 2011).

The exploration of unconventional resources has led to an escalating quantity of horizontal wells being drilled. This growth primarily stems from shale plays and is linked to hydraulic fracture stimulation or hydrofracking (Sondergeld et al., 2010). Figure 3.16 shows a schematic view of hydraulic fracturing. Horizontal wells typically have greater contact with the target formation, which is crucial for boosting the recovery from the source rock formations. In horizontal wells, raising the lateral length allows the well to remain within the desired geological zone, thereby enhancing its interaction with the formation.

Extended laterals are increasingly popular due to regulations on spacing, reduced environmental impact, and improved reservoir contact. Multiple wells can be drilled from a single pad using rigs with specific designs that have a skid arrangement, allowing for directional S-shaped wells to be drilled. This method reduces the footprint and optimizes surface facilities (Pilisi et al., 2010). Drilling technology, such as interactive 3D well-path designs and rotary steerable systems, enables the drilling of

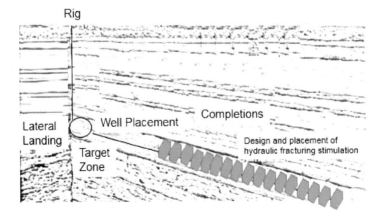

FIGURE 3.16   Schematic representation of hydraulic fracturing.

longer horizontal sections with improved precision. Extended horizontal well-bores have enhanced production rates, although they come with higher costs. The ideal length and the number of stages for these well-bores can be determined by maximizing the net present value (NPV) considering both production and expenses. An optimal lateral length and stage count can be determined by analyzing production history and different completion designs in reservoirs after several years of production. At times, the lateral length exceeds the coiled tubing's capacity. Additional concerns regarding drilling involve difficulties in the multi-well pad, maintaining position in the desired geological areas, ensuring optimal placement of the well (Kok et al., 2010) and stability of the well-bore, choosing appropriate drilling fluids, and preventing lost circulation, kicks, and blowouts (Pilisi et al., 2010; Zhang and Wieseneck, 2011). Subsurface formations exhibit heterogeneity, and the quality of shale reservoirs can vary significantly within a narrow section along the horizontal or vertical direction. Horizontal or directional wells pose challenges due to the irregularities in geological layers, which are frequently inclined, displaced, or uneven, causing difficulties in maintaining the lateral well-bore within the desired zone. For obtaining a detailed representation of the underground formation, real-time measurement techniques like measurement while drilling (MWD) or logging while drilling (LWD) may be used (Baihly et al., 2010).

Coupling 3D seismic surveys and LWD provides a more precise representation of the subsurface formation, enabling the accurate placement of lateral wells in the appropriate geological layers. For well placement, precise subsurface mapping functions as a global positioning system, facilitating optimal lateral steering within the desired zone. Drilling direction is crucial because breakdowns occur more frequently when the drilling of a horizontal well takes place in the maximum horizontal stress direction. Drilling in high-temperature-high-pressure reservoirs can pose challenges. Sophisticated drilling fluids enhance bore-hole integrity and minimize fluid loss, all while complying with environmental laws. Using the appropriate fluid can also facilitate faster and easier drilling. Most effective drilling fluids differ depending on the specific geological formation being drilled.

Hydraulic fracturing is typically necessary for tight reservoirs to be productive. In the case of shale reservoirs, the use of horizontal wells in combination with hydrofracking is often crucial for achieving commercial production (Hari et al., 2021; Warpinsiki et al., 2014; Waters et al., 2009). Hydrofracking is a method that involves inducing or repairing fractures in a rock formation by injecting proppants and fluids to enhance the flow of oil and gas from wells. Completing hydraulically fractured wells involves various processes and factors, such as completion method (cased and cemented- or open-hole), stage design, lateral landing, geometry of hydraulic fracture, frac-fluid, proppant characteristics, monitoring of hydraulic fractures, optimization of the treatment, reservoir volume stimulation, and analysis of production (Chong et al., 2010; King, 2010, 2014; Gao and Du, 2012; Agrawal et al., 2012; Manchanda and Sharma, 2014). An effective completion should involve creating, sustaining, and monitoring the networks of fractures and strategically positioning the perforation clusters. In this regard, it is necessary to comprehend the geological and geomechanical characteristics, natural fissures in the formation, reservoir heterogeneity, and optimized stimulation (Krishna et al., 2018). A precise Mechanical Earth Model

(MEM) is crucial for the development of shale reservoirs. The MEM should encompass the stress-field distributions and the mechanical properties' distribution in the underground formations, including the targeted reservoir zone as well as the layers above and below it (King, 2010, 2014). It is necessary because the stresses from the adjacent bounding formations affect the fracture height and length. The modeling process combines data from logging, core analysis, drilling, and completions to enhance comprehension of the mechanical characteristics and stress distribution across the stratigraphic column (Laik, 2018). An effective MEM facilitates the comprehension of how natural fractures are distributed and potentially reactivated during hydrofracking, allowing for a reliable forecast of hydraulic fracture characteristics such as length, growth in height, orientation, and aperture.

The completion design for unconventional reservoirs should consider both geomechanical aspects and reservoir characteristics (Ma and Holditch, 2015). Differentiating between completion quality and completion effort is crucial (Ma and La Pointe, 2011). "Completion quality" pertains to the characteristics of the rocks that determine whether completion is simple or complex, while "completion effort" refers to the different methods and tools utilized in the completion process. Completion efficiency is determined by combining completion quality and completion effort. Enhancing completion effectiveness involves considering the anisotropy in stress as well as other reservoir characteristics during hydrofracking and perforating operations. Uniformly fracturing the entire clusters belonging to a non-homogeneous zone cannot be ideal when the lateral well-bore extends into a heterogeneous reservoir. Sequential fracturing combined with a suitable fluid can enhance the effective length of fracture (Kraemer et al., 2014). Usually, a uniform design for perforation clusters is employed when the formation's heterogeneity is unknown. However, having proper comprehension of the lateral heterogeneity can assist in placing the perforation clusters in optimal locations, leading to increased hydrocarbon production from the perforated clusters. Key factors in completion design are fracture spacing, open-hole versus cased and cemented-hole configurations, tubular selection, fracturing point selection, sequential versus simultaneous fracking, fracture initiation points, frac-fluids, additives and proppants, the interaction of hydraulic fractures with natural fractures, and perforation strategies. These variables influence the fracture complexity, stimulated reservoir volume (SRV), and efficiency of a fracturing treatment. Key considerations in fracture design include: (1) in-situ stress state primarily influences the fracture orientation and azimuth; (2) hydraulic fractures become wider in rocks having lower Young's Modulus; (3) fracture length and containment are controlled by stress contrast; (4) planar hydraulic fractures are more likely in the absence of natural fractures; (5) transverse hydraulic fractures are expected when drilling a horizontal well in minimum horizontal stress direction; (6) the effective fracture length is constrained by the capability of the frac-fluid to carry the proppant over long distances; and (7) high injection rates typically enhance proppant transport and fracture length (Greff et al., 2014). Rock properties influence the vertical height and horizontal shape of fractures, which includes their length, width, and fracture complexity index. The height of hydraulic fractures is primarily determined by the in-situ stress state and can also be influenced by the viscosity of the frac fluid and the injection rate. When a fracture propagates to a point where the stress difference exceeds the pressure within the crack, it will come to a halt. A significant correlation exists between the net pressure and the width of the fracture. Higher injection rates generally lead to an increase in fracture width. The width of fractures is crucial because premature screen-outs can be led by narrower fracture width. An incremental rise in net pressure following the initial growth phase typically suggests the presence of a vertical fracture that is expanding in length. The effective length of fractures is typically constrained by the fracturing fluid's capacity to carry proppants over extended distances. Fracture height expansion and fracture width impact the propped-fracture half-length based on the treatment size. Higher injection rates generally enhance proppant transport and increase the effective fracture length. Fractures can vary in complexity and are influenced by stress anisotropy and the characteristics of natural fracture systems, which are crucial factors in determining the complexity of hydraulic fractures (Sayers and Calvez, 2010; Wang et al., 2015).

The interactions between hydraulic fractures and already-existing natural fractures are crucial in hydraulic fracture design. Additionally, if the horizontal minimum and maximum stresses are of similar magnitude, fractures are likely to propagate in multiple directions, resulting in a complex fracture network. Fracture complexity significantly affects production. However, while it enhances reservoir contact, it can make it difficult to establish a long-lasting proppant pack with adequate hydraulic continuity. Frac-fluids are substances added to the subsurface formation to enhance the flow of hydrocarbon fluids. Commonly used frac-fluids include water-based fluids (such as slickwater, linear gels, and crosslinked gels), oil-based fluids, foam-based fluids, and viscoelastic surfactant-based fluids (polymer-free fluids). The frac-fluid must maintain a consistent viscosity while being pumped and should be broken down toward the conclusion of the fracturing operation. The viscosity should be at a specific level to facilitate proppant transport and regulate the net fracture pressure. The frac-fluid's viscosity affects the geometry of the fracture. Mineralogy also influences the choice of frac-fluid. Formations with over 50% clay content pose challenges for completion, requiring careful fluid selection. Proppants in hydrofracking treatment are used to facilitate fracture opening, enabling the connection between the reservoir and wellbore for hydrocarbon flow (Hari et al., 2021). The type and quantity of proppants directly affect the recovery and the ability of fractures to sustain flow (Gao and Du, 2012; McKenna, 2014; Greff et al., 2014). Most shale reservoirs utilize sands as proppants, which must meet specific criteria such as uniform grain size, roundness, and crush resistance. Bauxite and ceramic-type proppants of smaller mesh sizes are utilized as well. The primary principle of proppant utilization and selection for hydraulic fracturing is to support the fractures without impeding the fluid flow between the reservoir and the well. Key factors to consider when choosing proppants are closure stress, flowing bottom-hole pressure (FBHP), and conductivity. Proper distribution of proppant within the networks of fractures is crucial for the efficiency of hydraulic fractures (Cipolla, 2009; Zhang and Wieseneck, 2011). Even for large hydraulic fractures, the effective length of the fracture is limited when the proppant accumulates in the primary planar fracture. Complex fracture growth often results in a low average proppant concentration, which may not significantly affect well performance (McKenna, 2014).

Evaluating and developing unconventional resources is more intricate compared to conventional resources. We need to employ a comprehensive strategy that includes assessing TOC and petroleum system's thermal maturity as well as evaluating petrophysical variables such as porosity, fluid saturation, permeability, and geomechanical properties. Designing drilling and completion processes and optimizing production should be grounded in the quality of the reservoir and completion to enhance the economic viability of unconventional resource development. There are numerous challenges in assessing unconventional resources and in maximizing their production. Understanding the impact of composite variables and spurious and spurious weak correlations can improve the design of completion processes to enhance production. With the expansion of the global economy, the exploration of unconventional resources is increasingly crucial, despite facing fluctuations. Advancements in technology now allow for extensive production from tight formations. Yet, numerous obstacles remain in the process of generating energy from unconventional resources. Optimizing current technologies and innovating new ones can harness these resources for societal benefit while reducing environmental harm.

3.7   GREEN ENERGY SYSTEMS: WAVE, WIND, AND SOLAR POWER

Energy is indispensable to humanity, as our society is unable to function without it. Energy has been the main driving force behind the Industrial Revolution, shaping the current world economy. Our energy supply system, reliant on fossil fuels for electricity generation and transportation, now poses a threat to humanity and the global ecosystem. Significant quantities of carbon dioxide released into the atmosphere daily are altering our climate. Scientific evidence has established that the enhanced greenhouse effect is causing a significant increase in air and water temperatures at a faster rate than in the pre-industrial era. It is resulting in the damage of natural habitats while the sheets of ice and sea ice are being lost. Ultimately, these lead to rising sea levels, drought, and extreme weather events. Decarbonizing our current energy supply system is a significant challenge for humanity. However, there are abundant clean and natural resources of energy available to meet the demands of the global population. The solar energy reaching Earth daily, in the form of light and heat, generates wind, sea waves, and thermoclines in the marine environment, surpassing mankind's energy requirements by a significant margin. The oceans, covering more than 71% of the Earth's surface, provide a significant source of clean energy through different renewable forms such as wave, offshore wind, tidal, and thermal and salinity gradients.

The main reasons coastal populations turn to exploiting the resources of energy available at sea are the limited land space in densely populated areas and the need for large onshore space to meet the energy demand using renewable energy sources. Utilizing renewable energy sources in the ocean presents various engineering obstacles that complicate the marine system design in comparison to those on land. Offshore structural integrity requirements are more stringent due to the harsh winds and rough waves experienced over a design lifespan of 20–30 years. The harsh marine environments (also corrosive) require more extreme measures to safeguard marine systems from material deterioration. Additional practical factors, like marine growth, ice in movement, scour caused by sea currents, and sea-vessel impact caused accidentally, necessitate a more rigorous engineering analysis in the design phase. Carrying out installation, operation, and maintenance processes in the offshore environment is more complicated than on land due to challenges such as limited site accessibility caused by severe weather conditions as well as the requirement for suitable port facilities and appropriate sea vessels.

Offshore wind energy involves producing electricity using wind turbines situated in water bodies, usually in the ocean. Offshore wind power has garnered considerable interest and expansion in recent years because of its numerous advantages compared to onshore wind energy. Offshore locations typically have greater and more reliable wind speeds, in comparison to onshore areas. Offshore wind turbines can produce increased electricity output. Offshore wind farms are typically situated at a distance from densely populated regions, minimizing visual disturbance and mitigating land use issues. Offshore wind farms can utilize bigger and more powerful turbines, in contrast to onshore installations. Increased turbine size results in greater wind energy capture and higher electricity production. Offshore locations provide more reliable wind conditions, resulting in more steady and foreseeable electricity production, in comparison to onshore wind farms. Offshore wind farms circumvent the need for valuable land resources, thus mitigating issues associated with land use and permitting obstacles encountered by onshore projects. Continuous progress in offshore wind turbine technology, such as floating platforms and installations in deeper waters, helps to increase the number of feasible offshore locations.

Offshore wind projects typically entail greater upfront expenses than onshore projects, particularly due to the costs related to building foundations in deep water. Maintaining offshore turbines can be difficult and expensive because of their remote location and limited accessibility during harsh weather. Offshore wind farms can affect marine ecosystems, so thorough environmental assessments are essential to reduce negative impacts. Establishing connections between offshore wind farms and the electrical grid necessitates the creation of dependable and effective transmission infrastructure. Acquiring the required permits and managing regulatory procedures for offshore wind projects can be time-consuming and intricate.

Global growth in offshore wind capacity is increasing, as numerous countries are investing in and expanding their offshore wind projects. Progress in floating wind turbine technology enables the installation of turbines in deeper waters, broadening the range of possible offshore sites. Continual attempts to lower expenses related to offshore wind, through advancements in technology and increased production levels, are enhancing the competitiveness of offshore wind energy. Offshore wind energy is essential for shifting to renewable energy sources and decreasing greenhouse gas emissions. The offshore wind sector is projected to grow due to technological advancements and changing regulations, making a substantial impact on the global energy supply.

Offshore wave energy involves capturing the energy produced by ocean waves to generate electricity. This renewable energy source can meet worldwide energy demands and decrease dependence on fossil fuels. Offshore wave energy is harnessed through Wave Energy Converters (WECs). WECs are engineered to transform the kinetic and potential energy of ocean waves into electrical energy. Offshore wave energy projects are usually situated in regions characterized by regular and substantial wave action, like open ocean settings or coastal areas with intense wave patterns. Ocean waves have a high energy density, which offers the possibility of generating a substantial amount of electricity. Ocean waves are more predictable than some other renewable energy sources, such as wind or solar, making wave energy a relatively stable and reliable power source. Offshore wave energy projects are typically situated in remote areas to reduce their visual impact on the surrounding landscape. Ocean waves provide a consistent source of energy for continuous electricity generation, unlike solar or wind power which is intermittent.

Offshore wave energy systems need to endure the severe marine environment, such as storms, corrosive saltwater, and difficult sea conditions. Offshore wave energy devices require specialized vessels and equipment for maintenance due to their remote location, which poses a challenge. Installing and using wave energy devices may impact the environment, potentially affecting marine ecosystems and navigation. Wave energy technology is currently in the early stages of development compared to other renewable energy sources, and its widespread deployment may encounter technical and economic obstacles. Multiple nations have carried out pilot initiatives to assess and showcase the practicality of offshore wave energy. Current research is concentrated on enhancing the efficiency, durability, and cost-effectiveness of wave energy technologies. Offshore wave energy can supplement other renewable sources, like offshore wind, in integrated energy systems. Offshore wave energy shows promise as a renewable resource, but its success and widespread use rely on addressing technical, economic, and environmental obstacles. Further, research and development, combined with supportive policy frameworks, will be essential for the advancement of offshore wave energy as a sustainable energy option. Offshore solar involves installing solar photovoltaic (PV) systems in bodies of water, like oceans, seas, or large lakes. Offshore solar projects have been suggested and created to address land limitations, optimize land use, and potentially improve the effectiveness of solar energy production, in contrast to traditional land-based solar installations. Offshore solar systems usually consist of solar panels mounted on floating platforms tethered to the seabed. The platforms can be moored in shallow or deep water, depending on the design and technology employed.

The solar panels in offshore solar installations are similar to those in traditional land-based systems. The floating platforms are engineered to endure the marine environment, including exposure to saltwater and wave action. Buoyant platforms are structures that float on the water's surface, with solar panels attached directly to them. Hybrid systems integrate floating solar panels with other offshore technologies like wind or wave to produce energy. Offshore solar enables the use of water bodies for generating solar energy, minimizing conflicts with land use for agriculture, or other activities. The water around the solar panels can provide a cooling effect, potentially enhancing the efficiency of solar cells that may be impacted by high temperatures. Offshore solar can be a viable alternative to land-based solar farms in densely populated coastal regions, alleviating space constraints. Offshore solar can be combined with other offshore renewable energy sources to create hybrid systems that optimize energy generation from the marine environment.

Creating buoyant structures that can endure marine environments, including waves, storms, and corrosion, is a complex technical task that demands specialized engineering expertise. Managing offshore solar installations is more difficult and costly compared to land-based systems. Offshore solar system deployment necessitates a thorough evaluation of potential environmental consequences, such as impacts on marine ecosystems, navigation, and local biodiversity. Offshore solar projects may have greater initial expenses than onshore installations, and their financial feasibility relies on variables like technology expenses, productivity improvements, and governmental backing.

Multiple pilot projects and research initiatives are currently being conducted worldwide to assess and showcase the practicality of offshore solar technologies. Current research is concentrated on enhancing the design, efficiency, and cost-effectiveness of offshore solar systems. Regulatory frameworks and policies are essential for the advancement and implementation of offshore solar projects in renewable energy. Offshore solar is a developing sector in renewable energy that may need to overcome technical obstacles and environmental issues and achieve economic viability compared to other renewable energy sources for widespread adoption. Further innovation, research, and collaboration among industry stakeholders will help expand offshore solar as a sustainable energy option.

Offshore energy resources are a vital part of the worldwide transition to cleaner and more sustainable energy. Advancements in technology, decreasing costs, and evolving regulatory frameworks are expected to make offshore wind, wave, and solar energy more important in meeting the world's increasing energy needs and reducing the effects of climate change. Collaboration among governments, industry stakeholders, and the research community is necessary to tackle challenges and make the most of opportunities presented by offshore energy resources.

4 Offshore Platforms

Summary

This chapter deals with the historical development of offshore platforms and their structural action under different environmental loads. The newly evolved structural forms and their discrete characteristics are also discussed. This chapter also gives the reader a good understanding of the structural action of different forms in the offshore platform, leading to the need to upgrade structural systems with time. An overview of the construction stages of offshore plants and their foundation systems helps the reader understand the construction methods and the allied complexities.

4.1 INTRODUCTION

Offshore structures are challenged to counteract oil resource depletion with discoveries (Adamas and Baltrop, 1991; Roren and Furnes, 1976; Sadehi, 2007). By 2010, increased drilling platforms induced the demand for offshore structures in the deep sea. Hence, the quest to research and develop deepwater structures has resulted in the recent advancement and thrust in this area (Adrezin et al., 1996, 1999). Expansion of the structures from shallow to deep waters makes the accessibility difficult, and hence, the structures demand higher deck areas consisting of additional space for third-party drilling equipment (Agarwal and Jain, 2002; Kjeldsen and Myrhaug, 1979; Sarpkaya, 1978; Sarpkaya and Isaacson, 1981). Specific challenges in arctic regions in shallow waters that arise due to low temperature, remoteness, ice conditions, ecosystem, and safety necessitate an adaptive design of offshore platforms addressing these factors (Chandrasekaran and Bhattacharyya, 2011). The development of offshore platforms depends on various factors; namely, i) structural geometry with a stable configuration; ii) easy to fabricate, install, and decommission; iii) low CAPEX; iv) early start of production; and v) high return on investment by increased and uninterrupted production. Newly generated structural forms do not have precedence in comparing and understanding their behavior and complexities. It is, therefore, essential to understand the structure's response and select the most suitable structure for the environment. It is one of the critical features of Front-End Engineering Design (FEED) (Devon and Jablokow, 2010). Figure 4.1 shows a drilling semisubmersible for deepwater drilling with vertical riser storage.

Offshore platforms fall under three major categories: (i) fixed platforms, (ii) compliant platforms, and (iii) floating platforms (API, 2005; Anagnostopoulos, 1982). Fixed-type platforms are further classified as jacket platform and gravity platform. Compliant platforms are further classified as jack-up rigs, guyed tower, articulated tower, tension leg platform, and spar. Floating platforms are classified as semisubmersibles; Floating Production Units (FPU); Floating Storage and Offloading (FSO); Floating Production, Storage, and Offloading (FPSO) Systems.

DOI: 10.1201/9781003497660-4

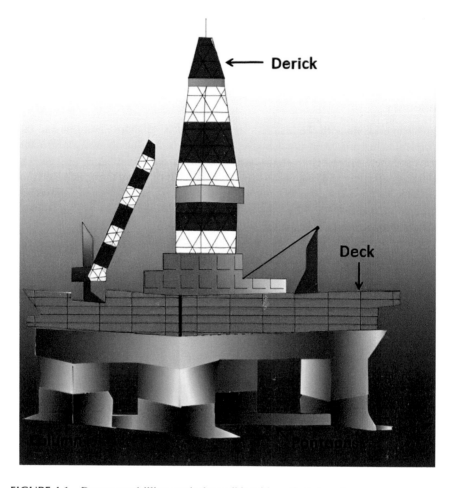

FIGURE 4.1   Deepwater drilling semisubmersible with vertical riser storage.

4.2   BOTTOM-SUPPORTED STRUCTURES

Energy is the driving force of the progress of civilization. Industrial advancements were first stoked by coal and then by oil and gas. Oil and gas are essential commodities in world trade. Oil exploration that initially started ashore has now moved to much deeper waters, owing to the paucity of resources in shallow waters (Bhattavharyya et al., 2010a; 2010b; Gusto MSC, 2010). Until now, over 20,000 offshore platforms of various kinds have been installed worldwide. Geologists and geophysicists search for potential oil reserves within the ground under the ocean sea floor, and engineers take the responsibility of transporting the oil from the offshore site to the shore location (Dawson, 1983). There are significant areas of operation from exploration to transportation of oil: (i) exploration, (ii) exploration drilling, (iii) development drilling, (iv) production operations, and (v) transportation (Chandrasekaran

FIGURE 4.1   (Continued).

and Bhattacharyya, 2011; Clauss et al., 1992; Clauss and Birk, 1996). Since the first offshore structure was constructed, more advanced design technologies emerged for building larger platforms that cater to deeper water requirements; each design is unique to the specific site (Ertas and Eskwaro-Osire, 1991). A precise classification of the offshore platform is difficult because of the large variety of parameters involved, such as functional aspects, geometric form, construction, and installation methods (Gerwick, 1986). However, the platforms are broadly classified based on the geometric configurations (Chandrasekaran and Yuvraj, 2013; Chandrasekaran and Nannaware, 2013; Chandrasekaran et al., 2013). Offshore installations are constructed for varied purposes: (i) exploratory and production drilling; (ii) preparing water or gas injection into the reservoir; (iii) processing oil and gas; (iv) cleaning the produced oil for disposal into the sea; and accommodation facilities (DOE-OG, 1985). They are not classified based on their functional use but on their geometric (structural) *form* (Sadehi, 2007; Sarpkaya and Isaacson, 1981).

As the platforms aim for greater water depths, their structural form changes significantly; alternatively, the same form cannot be used at a different depth. This means that the geometric evolution of the platform needs to be adaptive to counteract the environmental loads at the chosen water depths (Patel, 1989). Furthermore, the technological complexities faced by new offshore platforms, including analysis and design, topside details, construction, and installation, are not available in the open domain; they are protected and owned by the respective companies/agencies as part of their copyright. Because of such practices, knowledge of the complexities of designing offshore plants is not available to practicing young engineers. Hence, before gaining knowledge of FEED, it is necessary to understand different structural forms of offshore structures that have been successful. As it is well known that each platform is unique in many ways, learning about their structural configurations, limitations concerning the sea states and water depth, construction complexities, decommissioning issues, and their structural action will be an essential stage in the pre-FEED (Hsu, 1981). The present trend is to design and install offshore platforms in regions that are inaccessible and difficult to use the existing technologies (Anagnostopoulos, 1982). The structural form of every platform is derived mainly based on structural innovativeness but not based on functional advancement. Constructing existing platforms worldwide will impart decent knowledge to offshore engineers (Gerwick, 1986; Graff, 1981a, 1981b). Significant components of offshore platforms are:

(i) Superstructure: It consists of a deck and equipment for the platform's functioning.
(ii) Substructure: It supports the deck and transmits the load from the substructure to the foundation.
(iii) Foundation: It supports the substructure and superstructure and transmits the load to the seabed.
(iv) Mooring system: It is used for station keeping.

Offshore platforms are broadly classified either as bottom-supported or floating. Bottom-supported platforms can be classified as fixed or compliant-type structures; compliant means flexible (mobility). Compliance changes the dynamic behavior of such platforms. Floating structures are classified as neutrally buoyant type (e.g., semisubmersibles, FPSO, mono-column spars) and positively buoyant type (e.g., tension leg platforms). It is important to note that buoyancy is vital in floating-type offshore structures, as the classifications are based on buoyancy (Bea et al., 1999; Colwell and Basu, 2009; Fujino and Abe, 1993; Tait et al., 2008; Viet and Naghi, 2014; Muhari and Gupta, 1983; Munkejord, 1996). In bottom-supported structures, the base is fixed rigidly to the seabed. The structure attracts more force due to its rigidity, but the response to the wave loads is much less.

4.3 JACKET PLATFORMS

Fixed-type platforms are also called "template-type structures," which consist of the following:

- A jacket or a welded space frame, which is designed to facilitate pile driving and also acts as a lateral bracing for the piles
- Piles, which are permanently anchored to the seabed to resist the lateral and vertical loads that are transferred from the platform
- A superstructure consisting of the deck to support other operational activities.

Table 4.1 lists fixed offshore platforms constructed worldwide (Chandrasekaran and Jain, 2016). It is seen that a large bunch of platforms were initiated in the USA and Europe. A typical jacket platform (Bullwinkle platform) is shown in Figure 4.2. A typical jacket platform consists of a process, wellhead, riser, flare support, and living quarters.

TABLE 4.1
Fixed Offshore Platforms around the World

S. No.	Platform Name	Water Depth (m)	Location
North America			
1	East Breaks 110	213	USA
2	GB 236	209	USA
3	Corral	190	USA
4	EW910-Platform A	168	USA
5	Virgo	345	USA
6	Bud Lite	84	USA
7	Falcons' Nest production	119	USA
8	South Timbalier 301	101	USA
9	Ellen	81	USA
10	Elly	81	USA
11	Eureka	213	USA
12	Harmony	365	USA
13	Heritage	328	USA
14	Hondo	259	USA
15	Enchilada	215	USA
16	Salsa	211	USA
17	Cognac	312	USA
18	Pompano	393	USA
19	Bullwinkle	412	USA
20	Canyon Station	91	USA
21	Amberjack	314	USA
22	Bushwood	***	USA
23	Hebron	92	Canada
24	Hibernia	80	Canada
25	Alma	67	Canada

(Continued)

TABLE 4.1 (Continued)
Fixed Offshore Platforms around the World

S. No.	Platform Name	Water Depth (m)	Location
26	North Triumph	76	Canada
27	South Venture	23	Canada
28	Thebaud	30	Canada
29	Venture	23	Canada
30	KMZ	100	Mexico
South America			
1	Peregrino Wellhead A	120	Brazil
2	Hibiscus	158	Trin and Tobago
3	Poinsettia	158	Trin and Tobago
4	Dolphin	198	Trin and Tobago
5	Mahogany	87	Trin and Tobago
6	Savonette	88	Trin and Tobago
7	Albacora	***	Peru
Australia			
1	Reindeer	56	
2	Yolla	80	
3	West Tuna	***	
4	Stag	49	
5	Cliff Head	**	
6	Harriet Bravo	24	Australia
7	Blacktip	50	Australia
8	Bayu-Undan	80	Australia
9	Tiro Moana	102	New Zealand
10	Lago	200	Australia
11	Pluto	85	Australia
12	Wheatstone	200	Australia
13	Kupe	35	New Zealand
South America			
1	Peregrino Wellhead A	120	Brazil
2	Hibiscus	158	Trin and Tobago
3	Poinsettia	158	Trin and Tobago
4	Dolphin	198	Trin and Tobago
5	Mahogany	87	Trin and Tobago
6	Savonette	88	Trin and Tobago
7	Albacora	***	Peru
Asia			
1	QHD 32-6	20	China
2	Peng Lai	23	China
3	Mumbai High	61	India
4	KG-8 Wellhead	109	India

(Continued)

TABLE 4.1 (Continued)
Fixed Offshore Platforms around the World

S. No.	Platform Name	Water Depth (m)	Location
5	Bua Ban	***	Thailand
6	Bualuang	60	Thailand
7	Arthit	80	Thailand
8	Dai Huang Fixed wellhead	110	Vietnam
9	Ca Ngu Vang	56	Vietnam
10	Chim Sao	115	Vietnam
11	Oyong	45	Indonesia
12	Kambuna	40	Indonesia
13	Gajah Baru	***	Indonesia
14	Belumut	61	Malaysia
15	Bukha	90	Oman
16	West Bukha	90	Oman
17	Al Shaheen	70	Qatar
18	Dolphin	***	Qatar
19	Zakum Central Complex	24	UAE
20	Mubarak	61	UAE
21	Sakhalin I	***	Russia
22	Lunskoye A	48	Russia
23	Molikpaq	30	Russia
24	Piltun-Astokhskoye-B	30	Russia
25	LSP-1	13	Russia
26	LSP-2	13	Russia
27	Gunashli Drilling and Production	175	Azerbaijan
28	Central Azeri	120	Azerbaijan
29	Chirag PDQ	170	Azerbaijan
30	Chirag-1	120	Azerbaijan
31	East Azeri	150	Azerbaijan
32	West Azeri	118	Azerbaijan
33	Shah Deniz Production	105	Azerbaijan
Europe			
1	Brage	140	Norway
2	Oseberg A	100	Norway
3	Oseberg B	***	Norway
4	Oseberg C	***	Norway
5	Oseberg D	100	Norway
6	Oseberg South	100	Norway
7	Gullfaks A	138	Norway
8	Gullfacks B	143	Norway
9	Gullfacks C	143	Norway
10	Sleipner A	80	Norway
11	Sleipner B	**	Norway

(Continued)

TABLE 4.1 (Continued)
Fixed Offshore Platforms around the World

S. No.	Platform Name	Water Depth (m)	Location
12	Sleipner C	***	Norway
13	Valhall	70	Norway
14	Ekofisk Center	75	Norway
15	Varg Wellhead	84	Norway
16	Hyperlink	303	Norway
17	Draugen	250	Norway
18	Statfjord A	150	Norway
19	Statfjord B	***	Norway
20	Statfjord C	290	Norway
21	Beatrice Bravo	290	UK
22	Jacky	40	UK
23	Ula	40	UK
24	Inde AC	70	UK
25	Armada	23	UK
26	Auk A	88	UK
27	Fulmar A	84	UK
28	Clipper South	81	UK
29	Clair	24	UK
30	East Brae	140	UK
31	Lomond	113	UK
32	East Brae	86	UK
33	Alwyn North A	126	UK
34	Alwyn North B	126	UK
35	Cormorant Alpha	126	UK
36	Dunbar	145	UK
37	Nelson	***	UK
38	Schooner	100	UK
39	Andrew	117	UK
40	Forties Alpha	107	UK
41	Forties Bravo	107	UK
42	Forties Charlie	107	UK
43	Forties Delta	107	UK
44	Forties Echo	107	UK
45	Eider	159	UK
46	Elgin	93	UK
47	Elgin PUQ	93	UK
48	Franklin	93	UK
49	Babbage	42	UK
50	Alba North	158	UK
51	Alba South	138	UK

(Continued)

TABLE 4.1 (Continued)
Fixed Offshore Platforms around the World

S. No.	Platform Name	Water Depth (m)	Location
52	Judy	80	UK
53	Amethyst	30	UK
54	Buzzard	100	UK
55	Brigantine BG	29	UK
56	Brigantine BR	***	UK
57	Cecilie Wellhead	60	Denmark
58	Nini East	***	Denmark
59	Nini Wellhead	58	Denmark
60	South Arne	60	Denmark
61	Galata	34	Bulgaria

*Data not available.

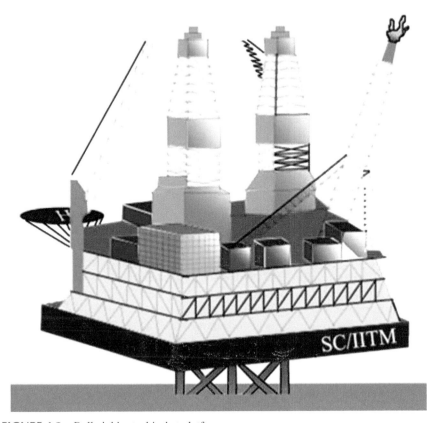

FIGURE 4.2 Bullwinkle steel jacket platform.

FIGURE 4.2   (Continued).

The platform leg braces the piles and serves as a guide for driving them; hence, it is called a "jacket structure." It is a truss-based system, which remains fixed to the seabed and transparent to wave loads. The platform is designed mainly for production purposes and is suitable up to a depth of 400 m. A jacket platform's different components are superstructure called "topsides." It is divided into deck and deck modules.

Deck: The upper part of the platform houses most of the equipment related to process, mechanical, electrical, piping, and instrumentation. It consists of three components—drilling deck, wellhead/production deck, and cellar deck. These decks are supported on a gridwork of girders, trusses, and columns. The deck structure is usually made up of Warren trusses.

Deck modules: The arrangement of deck modules into the available space on the deck structure is carefully planned during the preliminary design. Modules are fabricated with connecting pipe joints, including drilling equipment, production equipment, gas turbines, pumps, generating sets, compressors, gas flare stacks, helidecks, revolving cameras, survival crafts, and living quarters. It weighs about 40,000 tons.

Crane pedestal: A large structural tube that supports an offshore crane for lifting purposes. Crane pedestals also serve as diesel storage tanks, which are huge in diameter and capable of housing enough fuel.

Helideck: A raised level on the platform facilitates helicopter landing. Solar panels are mounted just below the helideck to reduce auxiliary power.

Flare boom: A long truss supports a vent or flare line that releases a part of hydrocarbon gas (about one-third of production) at a greater height.

(i)   Substructure:

Jacket: It is a supporting frame of the platform that supports the topsides and is generally submerged below the waterline. The wave loads mainly govern its design.

Transition piece: It is a structural member in the form of the cone that connects the topside and the jacket. It is a type of cup and cone joint. A cone-shaped design is preferable, as the leg size of the topsides is smaller in diameter than that of the jacket legs.

Legs: The jacket legs are usually battered in elevation to provide a more extensive base at the mudline and assist in overturning loads more effectively.

Braces: The main legs of the jacket are interconnected by horizontal and diagonal braces to improve their stability against lateral loads.

Buoyancy tanks: They provide adequate clearance between the seabed surface and the lowest point of the launched structure.

Conductors: These are long, hollow, vertical or curved tubes of 1 m diameter driven through guide rings below the mud line. The wells are drilled through conductor tubes positioned in an array, which can be reached by sliding the drilling derrick from location to location across the deck. Conductor framings are provided to support the long tube length and act as a lateral support guide to the conductors.

Risers: These are vertical tubes of 0.4 m diameter located within the jacket framework for pumping seawater into decks, heat exchange, and carrying the crude or partially processed oil/gas to another location for many other processing functions. Risers are generally clamped to the jacket structure.

Boat landings: Two boat landings at different levels are provided on each longitudinal side of the platform for embarking and disembarking.

Barge bumpers: These are provided on all jacket legs to accommodate landing/ unloading in a variety of sea conditions and to facilitate smooth berthing.

Riser guards: A riser guard protects the oil/gas-carrying risers. They are designed for accidental vessel impacts.

Launch truss: Sometimes, the jacket structures are immense and cannot be lifted, even with large cranes. Permanent structures like a launch truss are provided on one side of the jacket to facilitate loading onto the barge. If the jacket is designed for buoyancy, it is launched into the sea after reaching its position for a natural upend and leveling. When the jacket is launched, it floats due to buoyancy. The jacket legs are then sequentially flooded to make it upright and stand over the seabed before the piles are driven through the legs to fix it to the sea bed. The launch truss helps skid the jacket from the barge to the sea. It is provided to avoid damage to the main prominent truss members while launching. Rollers are provided to adjust the overlapping distance between the launch truss and the tower.

Joint Cans: These are provided to facilitate welding and expansion during the installation. The joint stress concentration is high and termed "hot spot stresses." At increased stresses, connections are generally damaged and need replacement. The number of joints in a leg member should be optimum as more extended members are susceptible to buckling, causing premature failure.

Piles: The topside loads are directly taken by piles and transferred to the soil. In this structure, the jacket structure takes the superstructure loads and moves to the piles through grouted sleeves. Piles support the platform in its worst in-service condition.

1. End bearing pile: The pile is resting on rocky strata. It transfers the load by bearing.
2. Skin friction pile: Longer than end bearing pile. It transfers load by friction.
3. Combined end bearing and skin friction pile.

Skirt piles: When the structure becomes heavier due to increased water depths, improved jacket configurations, known as skirt pile jackets, are required. The number of skirt piles to be established depends upon bottom soil conditions. Usually, they are placed in concentric circles around the base of the legs in rows of four, six, eight, or more. The skirt pile sleeves are incorporated in the jacket structure between the two lowest levels of horizontal bracing. The sleeves are sufficiently offset from the plane of the jacket side through pile guides. In deepwater applications, the skirt piles are clustered around enlarged corner legs to provide more excellent uplift resistance. Skirt piles are required when the soil is fragile and the existing number of piles formed by the geometry of the platform is not adequate.

The translational movement occurs in the platform due to improper fitting of fasteners and an increase in the flexibility of the piles in due course of time.

Mud mat: It is provided at the bottom of the jacket to give a suitable additional area to resist initial fluidization of the top layer of the seabed. They are provided to keep the platform stable and in a vertical upright position against the lateral forces before the piles are driven through the legs. They are usually made of heavy timbers or light steel plates and fastened across the corners of the jacket, immediately beneath the lowest level of horizontal bracing. It is similar to a large raft made up of timber. It helps the platform to sink deeper if the soil is too soft near the top layer of the sea bed. It provides adequate resistance to overturning.

Construction involves load-out, towing, launching, floating, upending, vertical position, piling, and deck mating. Dry towing is always preferred over wet towing, as the latter is severely affected by the sea state during installation. The most common procedure for installation is to let the jacket slide from the barge into the sea following a trajectory, which finally allows the jacket to float almost horizontally. Buoyancy tanks provide adequate clearance. From this naturally floating horizontal position, an upending operation is carried out to bring the structure to its vertical (desired) position. Then, the structure is flooded to set down in its final position and secured by driving piles into the seafloor. The advantages of offshore jacket platforms are as follows: (i) support large deck loads; (ii) possibility of being constructed in sections and transported; (iii) suitable for large field and long-term production (supports a large number of wells); (iv) piles used for foundation result in good stability; and (v) not influenced by seafloor scour (Jin et al., 2007) A few disadvantages are as follows: (i) cost increases exponentially with an increase in water depth; (ii) high initial and maintenance costs; (iii) not reusable; and (iv) steel structural members are subjected to corrosion, causing material degradation in due course of service life; (v) installation process is time-consuming and expensive.

4.4 GRAVITY PLATFORMS

In addition to steel jackets, concrete was prominently used to build some offshore structures. These structures are called gravity platforms or gravity-based structures (GBS). A gravity platform relies on the structure's weight to resist the encountered loads instead of piling (API-RP 2A, 2000; Hitchings et al., 1976; Hoeg, 1976; Hoeg et al., 1977). In regions where driving piles become difficult, structural forms are designed to lie on their weight to resist the environmental loads (Hove and Foss, 1974). Gravity platforms are large bottom-mounted reinforced concrete structures capable of supporting large topside loads during tow-out, which minimizes the hook-up work during installation (Leonard and Young, 1985). Additional large storage spaces for hydrocarbons add up to their advantage. Figure 4.3 shows a typical GBS. The major components of the gravity platform are as follows:

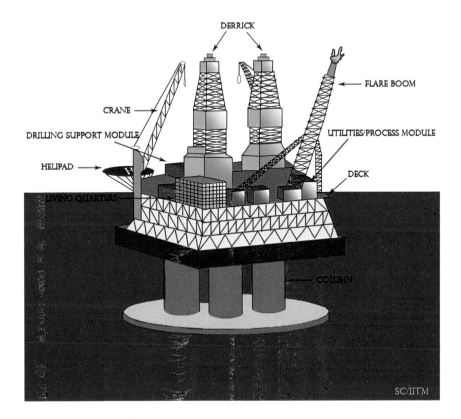

FIGURE 4.3   Typical GBS platform.

(i) *Foundation*: These structures have a foundation element that contributes significantly to the required weight and spreads over a large area of the seafloor to prevent failure due to overturning moments caused by lateral loads. The shape of the foundation shall be circular, square, hexagonal, or octagonal.

(ii) *Caissons*: They are hollow concrete structures at the platform's bottom. They provide the structure with natural buoyancy and are also used for storage. It enables the platform to float to field development location. After installation, void spaces are used as storage compartments for crude oil or filled with permanent iron or ballast. The height of the caissons is usually about one-third of the height of the platform.

(iii) *Steel Skirts*: The steel skirts are provided around the periphery of the main legs to improve stability against lateral movement of the platform against sliding; they also act as erosion-resistant members. In addition, they assist in grouting the caisson base. In addition to steel skirts, dowels are also provided, which extend about 4 m below the level of steel skirts. They help to prevent damage to steel skirts during touchdown operations.

(iv) *Modular Deck*: The deck supports drilling derrick, engine room, pipe rack, living quarters, processing equipment, and heliport. The prominent load-carrying members are plate girders, box girders, or trusses.

(v) *Cellar Deck*: They are used for placing the machinery and should be provided between the topside module and deck to improve the platform's stability. The platform's base is constructed in a dry dock, after which it is floated and moored in a sea harbor. The construction is then completed by slip-forming the large towers continuously until they are topped off. The structure is then ballasted (void spaces are flooded), and a steel prefabricated deck is floated over the structure and attached to its top (topside modules are mounted). When the caissons are filled with oil or ballast, the platform's center of gravity moves toward the bottom, making the platform sink. It is called "well sinking." The construction of gravity platforms requires deep harbors and deep tow-out channels.

The significant weight of the structure causes enormous soil erosion at the bottom. It also results in unequal settlement, due to which the structure tilts. Furthermore, tilting causes a shift in the platform's center of gravity, which creates an overturning moment at the base. The overturning moment increases with the lateral forces arising from waves, causing failure. In addition, a few geotechnical problems associated with GBS platforms cause large horizontal and vertical displacements. The geotechnical difficulties associated with GBS are:

- Sliding occurs due to the change in soil characteristics (friction) and lower wind and wave load resistance.
- Bearing capacity failure occurs due to overweight of the structure (punching failure) where the weight of the platform becomes more significant than the bearing capacity of the soil. The massive weight of the structure initiates local failure.

- Rocking occurs due to unequal settlement and causes a moment at the bottom of the platform.
- Liquefaction occurs due to poor soil conditions and saturation levels.

Their salient advantages include: (i) constructed onshore and transported; (ii) towed to the site of installation; (iii) quick installation by flooding; (iv) no exceptional foundation is required; and (v) use of traditional methods and labor for installation. Table 4.2 shows the list of gravity platforms constructed worldwide. These platforms are also known to be responsible for seabed scouring due to large foundations, causing severe environmental impact (Chandrasekaran, 2015b, 2015c, 2015d, 2016a, 2016b). Gravity platforms had severe limitations; namely, (i) not suitable for sites of poor soil conditions, as this would lead to significant settlement of foundation; (ii) long construction period, which thereby delays the early start of production;

TABLE 4.2
Gravity Platforms Constructed Worldwide

S. No.	Name of the Platform	Water Depth (m)
1	Ekofisk 1	70
2	Beryl A	119
3	Brent B	140
4	Frigg CDP1	98
5	Frigg TP 1	104
6	Frigg MCP01	94
7	Brent D	142
8	Statfjord A	145
9	Dunlin A	153
10	Frigg TCP2	103
11	Ninian	136
12	Brent C	141
13	Cormorant	149
14	Statfjord B	145
15	Maureen	95.6
16	Stafford C	145
17	Gulfaks A	133.4
18	GulfaksB	133.4
19	GulfaksC	214
20	Oseberg	100
21	Slebner	80
22	Oseberg North	100
23	Draugen	280
24	Heidrun	280
25	Troll	330

Courtesy PennWell Publishing Co.

and (iii) natural frequencies falling within the range of significant power of the input wave spectrum (Boaghe et al., 1998).

Gravity structures are constructed with reinforced cement concrete and consist of a large cellular base, surrounding several unbraced columns that extend upward from the base to support the deck and equipment above the water surface (Reddy and Arockiasamy, 1991). Gravity platforms consist of production risers as well as oil supply and discharge lines, contained in one of the columns; the corresponding piping system for the exchange of water is installed in another, and drilling takes place through the third column. This particular type is called a CONDEEP (concrete deep water) structure and was designed and constructed in Norway. The construction of gravity platforms requires deep harbors and deep tow-out channels. The floatation chambers are used as storage tanks, and platform stability is ensured through skirts. Steel gravity platforms exist off Nigeria, where rock close to the sea floor ruled out the possibility of using piles to fix the structures to the seabed. Figure 4.4 shows the

FIGURE 4.4 Hibernia gravity base structure.

FIGURE 4.4   (Continued).

Hibernia gravity base structure. The platform is a steel gravity base structure with a weight of 112,000 tons, a height of 241 m, and has steel skirts for penetration into the seabed.

The advantages of gravity platforms over jacket platforms are: (i) greater safety for people on board and topside; (ii) towing to the site with a deck is possible, which minimizes installation time and cost; (iii) low maintenance cost because concrete submerged in water will have lesser problems than that of steel structure; (iv) adjustable crude oil capacity; (v) capability to support large deck areas; (vi) risers are protected as they are placed inside the central shaft; and (vii) possible access to the seafloor from the cell compartment in the foundation, resulting in healthy monitoring.

4.5   JACK-UP RIGS

Jack-up (rigs) platforms are temporary structures meant for exploratory drilling. They are similar to barges with movable legs. They are mobile, as their hulls have the requisite floating characteristics to enable towing from site to site. When the legs are projecting upwards, the rig can be easily towed from one location to another. The jacking system provides an effective method to lower or raise the hull quickly. The legs are lattice, truss-type, and transparent to wave loads. When the jack-up is towed to the site for exploratory drilling, the legs will project upwards from the deck. On installation, the legs will be pushed inside the sea bed while the deck is lifted. Hence called a "Jack-up rig." After installation, one-third of the height of the leg should be left above the hull to maintain the platform's stability. The failure in the platform occurs during sailing when the legs are entirely above the hull due to an overturning moment caused by the wind load and by spud can pull off. The latter may cause severe damage to the drill pipes and risers, but the system will remain floating. The spud can foundation is not an ideal hinged joint. It offers partial fixity to the structure, which may also fail under bending. The jack-up rigs can work under harsh environments of wave heights up to 24 m and wind speeds exceeding 100 knots.

The components of the jack-up rig include the following:

Derrick: The derrick has the shape of a frustum or cone, and tubular members are used for the construction. The derrick moves over the rails in the hull to house a more significant number of wells. This movement will not affect the functionality of the platform.

Draw works: The assembly of pipes used for drilling. The drilling occurs through the center pipe, and the oil comes out under pressure through the circumferential pipes. The center pipe is also loaded with drilling mud, which acts as a counterweight to balance the bottom pressure and avoids the movement of the drill bit vertically upwards during the drilling process.

Drill floor: The floor where the drilling derrick is located.

Drill pipe: It is used for oil extraction.

Drill string: The sharp wedge-shaped tip of the drill pipe used for drilling is called drill string.

Cantilever: The derrick moves over the cantilever projection. It is called "Offset drilling," by which the hull's center of gravity in the trapezoidal cross section will be shifted towards the hull's center. Since drilling is performed on the cantilever projection, there is the most minor disturbance to the deck and other processes.

Legs: The legs are lattice truss-type. The length of the legs should be more significant than 1.4 times the water depth. Lacing and battens on the legs provide additional stability.

Hull: It houses all the facilities, which include living quarters and a helideck.

A spud can is a shallow conical underside footing used for placing the leg on the seafloor. When the spud can is pushed into the sea bed under more significant pressure, a partial vacuum will be created inside the spud can. Due to suction force, the clay fills the void space, and it is fixed to the seabed. Once fixed, a high pull-out force is required for extraction. Diameter varies from 2.5 to 4 m, while immersion depth is about 1–2 m. To remove the spud can, soil in the circumferential area should be excavated, enabling soil inside the spud can to flow out.

Moonpool: It is the vent provided for the passage of the drilling rig in the hull. Advantages of the platform include (i) high mobility; (ii) low cost and efficient; iii) easy fabrication and repair; (iv) easy decommissioning; and (v) simple construction.

These platforms also have some severe limitations, such as (i) being suitable only for shallow depth, (ii) being subjected to sea bed scouring, which leads to differential settlement, and (iii) not being suitable for rocky stratum. The name *jack-up* is assigned as the legs will be pulled up while transported from one site to another. On reaching the installation site, legs will be driven into the seabed for better stability. Jack-up rigs have significant mobility, but the geometric configuration is comparable to a fixed-base structure. Figure 4.5 shows a schematic view of a typical jack-up rig. The rig will be preloaded to test whether the foundation of the legs has reached the desired level of lateral stability. Once the preloaded test is completed, the deck is further lifted to obtain the desired air gap that is required for safe operation. The air gap is provided to ensure that the deck does not interfere with the tides during operation. The foundation of a jack-up rig is vital to ensure its stability against lateral loads caused by waves and wind. Lattice tower-type legs of jack-up rigs are supported on a spud can (Koo et al., 2004). It is a shallow, conical-shaped cup, which is provided on

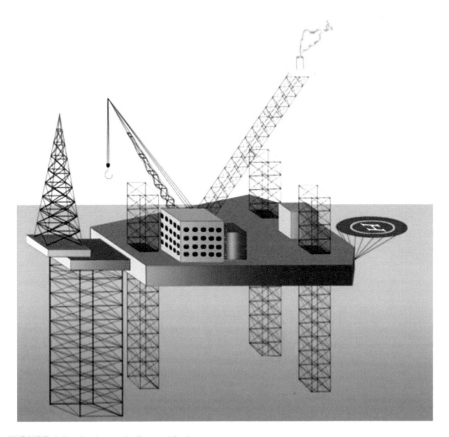

FIGURE 4.5   Jack-up platforms (rigs).

FIGURE 4.5 (Continued)

the underside of the footing of each leg. Figure 4.6 shows a schematic view of a spud used in the foundation of each of the legs of jack-up rigs. Spud cans are suitable for stiff, clay, and sand but not for rocks.

4.6 COMPLIANT STRUCTURES

To overcome the foregoing negative factors, one should design a structural form, which should attract fewer forces and remain flexible to withstand the cyclic forces. The structural action and the form are corrected based on the "mistakes" learned from the fixed type platforms. This is a special kind of reverse engineering that makes offshore platforms unique. This leads to continuous improvement from one platform to the other. Hence, FEED is on a constant update as new structural forms are being tried for oil and gas exploration in deep and ultra-deep waters (Chandrasekaran, 2014a; Stansberg et al., 2002, 2004). Fixed-type offshore structures became increasingly expensive and difficult to install in greater water depths. Hence, the modified

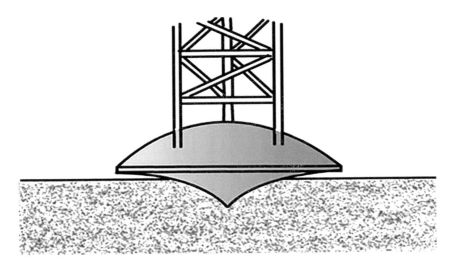

FIGURE 4.6   Spud can (Chandarsekaran and Jain, 2016).

design concept evolved for structures in water depths beyond 500 m. A compliant offshore tower is similar to that of a traditional platform, which extends from the surface to the sea bottom and is transparent to waves. A compliant tower is designed to flex with the forces of waves, wind, and current. Classification under compliant structure includes those structures that extend to the ocean bottom and are anchored directly to the seafloor by piles and/or guidelines. Guyed towers, articulated towers, and tension leg platforms fall under this category (Chandrasekaran et al., 2010a, 2010b; Helvacioglu and Incecik, 2004; Henselwood and Phillips, 2006; Hitchings et al., 1976). The structural action of complaint platforms is significantly different from that of the fixed ones, as they resist lateral loads not by their weight but by the relative movement. In fact, instead of resisting the lateral loads, the structural geometry enables the platform to move in line with the wave forces (Hoeg, 1976; Hoeg and Tong, 1977). To facilitate the production operation, they are position-restrained by cables/tethers or guy wires. By attaching the wires to the complaint tower, the majority of the lateral loads will be counteracted by the horizontal component of the tension in the cables; the vertical component adds to the weight and improves stability (Chakrabarti, 1994; Dawson, 1983; Boom et al., 1984).

4.6.1   GUYED TOWER

A guyed tower is a slender structure made up of truss members that rest on the ocean floor and are held in place by a symmetric array of catenary guy lines. The foundation of the tower is supported with the help of a spud arrangement, which is similar to the inverted cone placed under suction. The structural action of the guyed tower makes its innovation more interesting, which is one of the successful improvements in the structural form in the offshore structural design. The upper part of the guy wire is a lead cable, which acts as a stiff spring in moderate seas. The lower portion is a heavy chain, which is attached with clump weights called touch-down points. The clump weights are provided to drag down the cables to the seabed and drag anchors are provided to avoid lifting of clump weight due to scour. Under normal operating conditions, the weights will remain at the bottom, and the tower deck motion will be nearly insignificant. However, during a severe storm, the weights on the storm-ward side will lift off the bottom, softening the guying system and permitting the tower and guying system to absorb the large wave loads. Since the guy lines are attached to the tower below the mean water level close to the center of applied environmental forces, large overturning moments will not be transmitted through the structure to the base. This feature has allowed the tower to be designed with a constant square cross-section along its length, reducing the structural steel weight as compared with that of a conventional platform (Moe and Verley, 1980). Guyed towers are considered as pinned-pinned beams in analysis. In guy lines, the initial axial force is zero. It invokes the force only on larger displacement. The length of the guy wire is 2.5–3 times the water depth. The guy wires are provided circumferentially around the tower (possibly symmetrical), and they are connected to the top of the deck by "top tension risers." The tower is supported by a spud can, which is an inverted cone-shaped foundation system that resists the lateral force by suction pressure. The failure may occur because spud can pull off either a fair lead or a touchdown point.

Exxon 1983 installed the first guyed tower, named Lena Guyed Tower, in the Mississippi Canyon Block, at a 300 m water depth. Figure 4.7 shows the schematic view of the Lena tower. Though the structural form resembles a jacket structure, it is compliant and is moored by catenary anchor lines. The tower has a natural period of 28 seconds in sway mode while bending and torsion modes have a period of 3.9 and

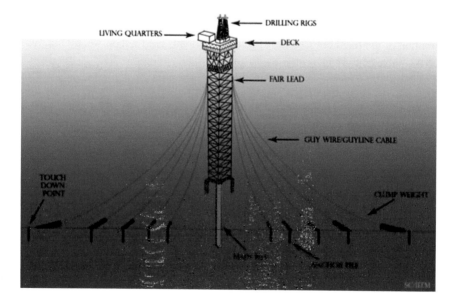

FIGURE 4.7   Lena Guyed tower.

FIGURE 4.7 (Continued)

5.7 seconds, respectively. The tower consists of 12 buoyancy tanks with a diameter of 6 m and a length of about 35 m. Around 20 guy lines are attached to the tower with clump weights of about 180 tons to facilitate the holding of the tower in position. The advantages of guyed towers are (i) low cost (lower than steel jackets); (ii) good stability as guy lines and clump weights improve restoring force; and (iii) possible reuse. The disadvantages are as follows: (i) high maintenance costs; (ii) applicable to small fields only; (iii) exponential increase in cost with increase in water depth; and (iv) difficult mooring. These factors intuited further innovation in the platform geometry, which resulted in articulated towers (Choi and Lou Jack, 1991).

4.6.2 ARTICULATED TOWER

One of the earliest compliant structures that started in relatively shallow waters and slowly moved into deep water is the articulated tower. An articulated tower is an upright tower that is hinged at its base with a universal joint, which enables free rotation about the base. When there was a need to improve the structural form from fixed to compliant, researchers thought of both modes of compliancy; namely, (i) rotational and (ii) translational. Enabling large translational motion could make the platform accessible from position-restrained, and hence, rotational compliance was attempted. In such geometric forms, essential to note that the design introduces a single-point failure deliberately, which is the universal joint (Helvacioglu and Incecik, 2004).

The tower is ballasted near the universal joint and has a large buoyancy tank at the free surface to provide a large restoring force (moment). The buoyancy chamber acts as a large container to store crude oil, and the lower ballast chamber is filled with permanent iron ore or ballast, which is used for shifting the center of gravity toward the bottom to provide more stability to the tower. Due to the lateral force, the structure tilts, which causes a shift of buoyancy (variable submergence). Depending upon location and size, the buoyancy chamber keeps on giving extra force in an upward direction, which will cause a couple in anticlockwise directions. This will restore the platform to its normal position. This is achieved by the dynamic change in the water plane area or variable submergence of the member (Wilson, 1984). In addition, the compliance of the articulated tower avoids the concentration of high overturning moments and the resulting stress (Choi and Lou Jack, 1991; Clauss and Birk, 1996; Clauss et al., 1992; Islam and Ahmad, 2003).

The tower extends above the free surface and accommodates a deck and a fluid swivel. In deeper water, it is often advantageous to introduce double articulation, the second one being at a mid-depth. The provision of more articulation reduces the bending moment along the tower. Fatigue is an essential criterion for this type of system design, as the universal joints are likely to fail under fatigue loads. The period varies from 40 to 90 seconds, which results in a dynamic amplification factor lesser than that of the fixed platforms. The advantages of articulated towers are as follows: (i) low cost; ii) large restoring moments due to the high center of buoyancy; and (iii) protection of risers by the tower. There are a few disadvantages: (i) suitable only for shallow water as the tower shows greater oscillations for increased water depth; (ii) cannot operate in bad weather; (iii) limited to small fields; and (iv) fatigue of universal joint leads to a single-point failure (Chandrasekaran and Pannerselvam, 2009; Chandrasekaran et al., 2010a, 2010b). They are used only for anchoring, storage, and repairing works and not as permanent production structures hence called single anchor leg mooring systems (SALM). Figure 4.8 shows a typical articulated tower supporting the storage activities of a vessel.

In both of the foregoing structural forms of complaint towers—namely, guyed tower and articulated tower—it can be seen that the tower extends through the water depth, making it expensive for deep waters. Therefore, successive structural forms are intuited toward the basic concept of not extending the tower to the full water depth but only retaining it near the free surface level as far as possible. In such kinds of structural geometry, it is inevitable to make the platform weight dominant. To improve the installing features and decommissioning procedures, the geometry is attempted to be buoyancy dominant instead of weight dominant. Buoyancy force exceeds the weight by manifold. While this enabled easy fabrication and installation, it also demanded skilled labor and high expertise for the installation and commissioning of such platforms. The evolved structural geometry is tension leg platforms (Vannucci, 1996; Yan et al., 2009; Yoneya and Yoshida, 1982; Demirbilek, 1990).

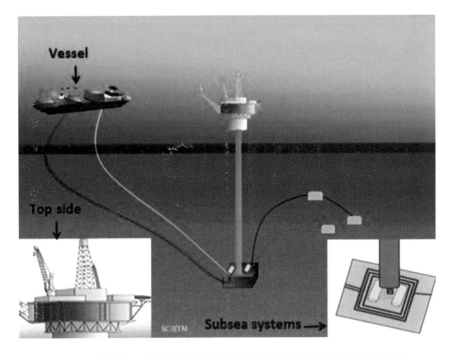

FIGURE 4.8 Articulated tower.

4.6.3 TENSION LEG PLATFORM

A tension leg platform (TLP) is a vertically moored compliant platform; (Basim et al., 1996; Bearman and Russell, 1996; Chen et al., 2006; Haritos, 1985; Simiu and Leigh, 1984; Simos and Pesce, 1997). Figure 4.9 shows a typical TLP, highlighting its various components. Taut mooring lines vertically moor the floating platform with its excess buoyancy; they are called tendons or tethers (Kim et al., 1994, 2007). The structure is vertically restrained, while it is compliant in the horizontal direction permitting surge, sway, and yaw motions (API RP 2T, 1997; Arnott et al., 1997; Bar-Avi, 1990; Bar-Avi and Benaroya, 1999; Chandrasekaran et al., 2006a; Chandrasekaran and Gaurav, 2008; Chandrasekaran and Jain, 2002a, 2002b; Kobayashi et al., 1987; Kurian et al., 2008; Kurian et al., 2008). The structural action resulted in low vertical force in rough seas, which is the critical design factor (Chandrasekaran and Jain, 2002a, 2002b; Rijken and Niedzwecki, 1991; Roitman et al., 1992; Booton et al., 1987). Columns and pontoons in TLP are constructed with tubular members due to which the buoyancy force exceeds the weight of the platform (Kareem and Datton, 1982; Kareem and Zhao, 1994; Karimirad et al., 2011; Logan et al., 1996;

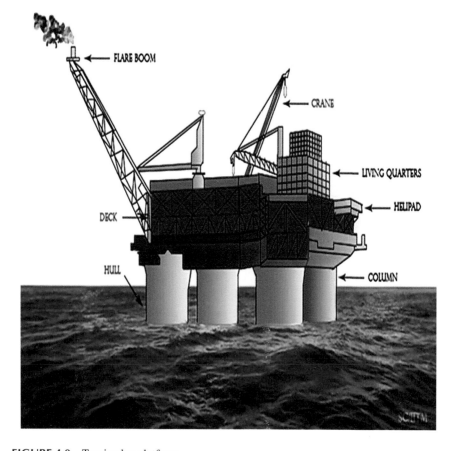

FIGURE 4.9   Tension leg platform.

Low, 2009; Marthinsen et al., 1992). The excess buoyancy created is balanced by the pretension in the taut moorings. Substantial pretension is required to prevent the tendons from falling slack even in the deepest trough, which is achieved by increasing the free-floating draft (Chandrasekaran et al., 2006a, 2006b; Basim et al., 1996; Kawanishi et al., 1987, 1993).

As the requirement of pretension is too high, pretension cannot be imposed in tethers by any mechanical means. During commissioning, void chambers (columns and pontoon members) are filled with ballast water to increase the weight; this slackens the tendons (Ker and Lee, 2002; Lee and Juang, 2012; Lee and Wang, 2000; Lee et al., 1999; Leonard and Young, 1985). After tendons are securely fastened to the foundation in the seabed, de-ballasting is carried out to impose necessary pretension in the tendons (Masciola and Nahon, 2008).

Under the static equation of equilibrium, the following relationship holds good:

$$W + 4T_0 = F_B \tag{4.1}$$

where W is the weight of the platform, T_0 is the initial pretension in tethers (about 20% of the total weight), and F_B is the buoyancy force, acting upwards. A pinned connection is provided between the deck and the mooring lines to hold down the platform in the desired position. Due to lateral forces, the platform moves along the wave direction. Horizontal movement is called offset. Due to horizontal movement, the platform also tends to have an increased volume of members. Thus, the platform will undergo a set-down effect. The lateral movement increases the tension in the tethers. The horizontal component of tensile force counteracts the wave action and the vertical component increases the weight, which will balance the additional weight imposed by set-down.

TLP is a hybrid structure with two groups of natural periods (Rijken and Niedzwecki, 1991; Roitman et al., 1992; Yan et al., 2009; Yashima, 1976; Yoneka and Yoshida, 1982; Yoshida et al., 1984; Younis et al., 2001; Zeng et al., 2007). Typical natural periods of the TLP are kept away from the range of wave excitation periods and typically for TLP resonance periods of 132 s (surge/sway) and 92 s (yaw) as well as 3.1 s (heave) and 3.5 s (pitch/roll), which are achieved through proper design (Nordgren, 1987; Chandrasekaran and Jain, 2004; Chandrasekaran et al., 2004, 2006b, 2007a, 2007b, 2007c; Boom et al., 1984). The main challenge for the TLP designers is to keep the natural periods in heave and pitch below the range of significant wave energy, which is achieved by an improved structural form (Basim et al., 1996; Kobayashi et al., 1987; Low, 2009; Chandrasekaran et al., 2012; Demerbilek, 1990; Devon and Jablokow, 2010; DNV, 1982, 2005, 2010; DOE-OG, 1985; Muren et al., 1996; Tabeshpour et al., 2006; Tabeshpour, 2013; Taflandis et al., 2008, 2009). The failure may occur either due to tether pull-out or fatigue effect on the tethers (Iwaski, 1981; Jain, 1997; Jeffreys and Patel, 1982; Jefferys and Rainey, 1994; Kam and Dover, 1988, 1989; Vannucci, 1996; Vickery, 1995, 1990). TLP mechanism preserves many of the operational advantages of a fixed platform while reducing the cost of production in water depths up to about 1500 m (Iwaski, 1981; Haritos, 1985; Chandrasekaran et al., 2004, 2007b, 2007c, 2013, 2006b, 2015a; Chandrasekaran and Jain, 2004; Donley and Spanos, 1991; Dyrbe and Hansen, 1997; Murray and Mercier, 1996). Its production and maintenance operations are similar to those of fixed platforms.

TLPs are weight-sensitive but have limitations in accommodating heavy payloads (Tabeshpour et al., 2006; Yoshida et al., 1984; Ertas and Lee, 1989; Chandrasekaran and Yuvraj, 2013; Gie and De Boom, 1981). Usually, a TLP is fabricated and towed to an offshore well site wherein the tendons are already installed on a prepared seabed (Mekha et al., 1996; Mercier, 1982, 1997; Meyerhof, 1976; Nordgren, 1987; Patel, 1989; Patel and Park, 1991, 1995; Patel and Witz, 1991; Patel and Lynch, 1983). Then, the TLP is ballasted down so that the tendons may be attached to the TLP at its four corners. The mode of transportation of TLP allows the deck to be joined to the TLP at the dockside before the hull is taken offshore (Bar-Avi, 1999; Adrezin et al., 1996; Chandrasekaran, 2007, 2013a, 2013c; El-Gamal et al., 2013; Ertas and Lee, 1989; Ney et al., 1992; Niedzwecki et al., 2000; Niedzwecki and Huston, 1992).

The advantages of TLPS are as follows: (i) mobile and reusable; (ii) stable as the platform has minimal vertical motion; (iii) low increase in cost with increase in water depth; (iv) deepwater capability; and (v) low maintenance cost (Perrettand and Webb, 1980; Perryman et al., 1995). A few disadvantages are: (i) high initial cost; (ii) high subsea cost; (iii) fatigue of tension legs; (iv) difficult maintenance of subsea systems; and (v) little or no storage. Table 4.3 highlights a few of the TLPs constructed worldwide (Chandrasekaran and Jain, 2016).

TABLE 4.3
Tension Leg Platforms Constructed Worldwide

USA

1	Shenzi	1333	USA
2	Auger	872	USA
3	Matterhorn	869	USA
4	Mars	896	USA
5	Marlin	986	USA
6	Brutus	1036	USA
7	Magnolia	1433	USA
8	Marco Polo	1311	USA
9	Ram Powell	980	USA
10	Prince	454	USA
11	Neptune	1295	USA
12	Ursa	1222	USA
13	Morpeth	518	USA
14	Allegheny	1005	USA
15	Jolliet	542	USA

Europe

1	Snorre A	350	Norway
2	Heidrun	351	Norway

Africa

1	Okumu/Ebano	500	Equatorial Guniea
2	Oveng	280	Equatorial Guniea

4.7 FLOATING PLATFORMS

Offshore platforms, which do not have any fixed base or permanent position-re-straint systems, are classified as floating platforms. While in operation, they are position-restrained either using a dynamic positioning system or mooring systems. Semisubmersibles, FPSO systems, FPUs, FSO systems, and spar platforms are grouped under this category (API, 2005).

4.7.1 SEMISUBMERSIBLES

Semisubmersible marine structures are well-known in the oil and gas industries and belong to the category of neutrally buoyant structures. These structures are typically moveable only by towing. These semisubmersibles have a relatively low transit draft, with a large water plane area, which allows them to be floated to a stationing location. On location, it is ballasted, usually by seawater, to assume a relatively deep draft or semi-submerged condition, with a smaller water plane area, for operation. Semisubmersible platforms have the principal characteristic of remaining in a substantially stable position and have minimal motions in all degrees of freedom due to environmental forces such as the wind, waves, and currents. The main parts of the semisubmersibles are the pontoons, columns, deck, and mooring lines. The columns bridge the deck and the pontoons—i.e., the deck is supported by columns. Flotation of semisubmersibles is accomplished with pontoons. The pontoons provide a relatively large water plane area, as is desirable for transit. When submerged for stationing and operations, the columns connecting the pontoons to the upper deck present a lower water plane area, thereby attracting fewer wave loads and thus reducing the motions. Generally, dynamic position-keeping systems (DPS) are deployed to hold the semisubmersibles in position during production and drilling (Patel and Witz, 1991). Figure 4.10 shows a typical semisubmersible.

The advantages of semisubmersibles are as follows: (i) mobility with high transit speed (10 knots); (ii) stability as they show minimal response to wave action; and (iii) large deck area.

A few disadvantages are (i) high initial and operating costs; (ii) limited deck load (low reserve buoyancy); (iii) structural fatigue; (iv) the cost to move large distances; (v) availability of limited dry-docking facilities; and (vi) difficulty in handling mooring systems and land BOP stack and riser in rough seas. Table 4.4 shows a list of semisubmersibles commissioned worldwide (Chandrasekaran and Jain, 2016).

4.7.2 FLOATING PRODUCTION, STORAGE, AND OFFLOADING (FPSO) PLATFORM

FPSO is an acronym for Floating Production, Storage, and Offloading Systems. Offloading of the crude oil is usually to a shuttle tanker. Typically, converted or newly built tankers are custom-made for the production and storage of hydrocarbons. These stored hydrocarbons are subsequently transported by other vessels to terminals or deepwater ports. The design variants of FPSO are FPS and FSO. FPS is an acronym for Floating Production Systems devoid of storage facility. Now, it is

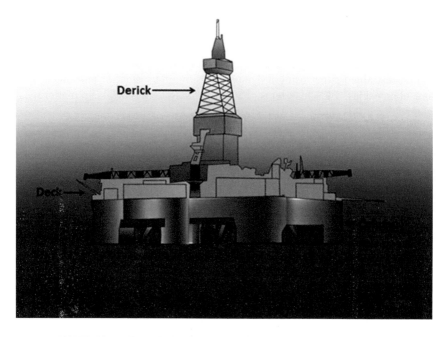

FIGURE 4.10 Typical semisubmersible.

TABLE 4.4
Semisubmersibles Commissioned Worldwide

S. No.	Platform Name	Location	Water Depth (m)	Year of Commissioning
Europe				
1	Argyll FPU	UK	150	1975
2	Buchan A	UK	160	1981
3	Deep Sea Pioneer FPU	UK	150	1984
4	Balmoral FPV	UK	150	1986
5	AH001	UK	140	1989
7	Janice A	UK	80	1999
8	Northern producer FPF	UK	350	2009
9	Asgard B	Norway	320	2000
10	Kristin FPU	Norway	320	2005
11	Gjoa	Norway	360	2010
12	Veslefrikk B	Norway	175	1989
13	Troll B FPU	Norway	339	1995
14	Njord A	Norway	330	1997
15	Visund	Norway	335	1999
16	Troll C FPU	Norway	339	1999
17	Snorre B FPDU	Norway	350	2001
USA				
1	Innovator	North America	914	1996
2	Nakika	North America	969	2003
3	Atlantis	North America	2156	2006
4	ATP Innovator	North America	914	2006
5	Thunder Horse	North America	1849	2008
6	Blind Faith	North America	1980	2008
7	Thunder Hawk	North America	1740	2009
8	P-09	Brazil	230	1983
9	P-15	Brazil	243	1983
10	P-12	Brazil	100	1984
11	P-21	Brazil	112	1984
12	P-22	Brazil	114	1986
13	P-07	Brazil	207	1988
14	P-20	Brazil	625	1992
15	P-08	Brazil	423	1993
16	P-13	Brazil	625	1993
17	P-14	Brazil	195	1993
18	P-18	Brazil	910	1994
19	P-25	Brazil	252	1996
20	P-27	Brazil	533	1996
21	P-19	Brazil	770	1997
22	P-26	Brazil	515	2000

(Continued)

TABLE 4.4 (Continued)
Semisubmersibles Commissioned Worldwide

23	P-36	Brazil Campos basin	1360	2000
24	P-51	Brazil Campos basin	1255	2001
25	SS-11	Brazil	145	2003
26	P-40	Brazil	1080	2004
27	P-52	Brazil	1795	2007
28	P-56	Brazil	1700	2010
29	P-55	Brazil	1707	2012
Asia				
1	Tahara	Indian Ocean	39	1997
2	Nan Hia Tiao Zhan	South China sea	300	1995
3	Gumusut Kakap	Malaysia	1220	2011

a universal term to refer to all production facilities that float rather than structurally supported by the seafloor, and typical examples include TLPs, spars, semisubmersibles, shipshape vessels, etc.

FSO is an acronym for Floating, Storage, and Offloading system. Like the FPSO, these are typically converted or newly built tankers, and they differ from the FPSO by not incorporating the processing equipment for production; the liquids are stored for shipment to another location for processing. Offloading indicates the transfer of produced hydrocarbons from an off-shore facility into shuttle tankers or barges for transport to terminals or deepwater ports. Figure 4.11 shows a typical FPSO used in oil and gas exploration. An FPSO relies on subsea technology for the production of hydrocarbons and typically involves pipeline export of produced gas with shuttle tanker (offloading) transport of produced liquids (Leffler et al., 2011). FPSOs are usually ship-shaped structures and are relatively insensitive to water depth. Mooring systems of FPSOs are classified as "permanent mooring" or "turret mooring." The majority of FPSOs deployed worldwide are permanently moored—i.e., the FPSOs with their moorings and riser systems are capable of withstanding extreme storms in the field. On the other hand, disconnectable FPSOs have attracted more attention recently. They are typically turret moored. Disconnectable turret is designed for FPSO to be able to disconnect to avoid specific extreme environments.

Salient advantages of the FPSOs are as follows: (i) low cost; (ii) mobile and reusable; (iii) reduced lead time; (iv) quick disconnecting capability, which can be helpful in iceberg-prone areas; (v) little infrastructure required; and (vi) turret mooring system enables FPS (converted ship type) to head into the wind/waves reducing their effect. A few disadvantages are (i) limited to small fields; (ii) low deck load capacity; (iii) damage to risers due to motion; (iv) poor stability in rough seas; and (v) little oil storage capabilities. Table 4.5 shows details of FPSOs commissioned worldwide (Chandrasekaran and Jain, 2016).

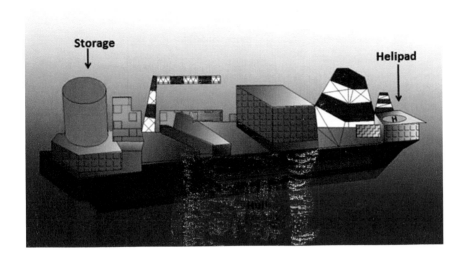

FIGURE 4.11   FPSO platform.

FIGURE 4.11   (Continued).

TABLE 4.5
Details of FPSO Commissioned Worldwide

S. No	Platform	Water Depth (m)	Location
Australia			
1	Maersk Ngujima-Yin	400	Australia
2	Stybarrow Venture	825	Australia
3	Pyrenees Venture	200	Australia
4	Glass Dowr	344	Australia
5	Front Puffin	110	Australia
6	Crystal Ocean	170	Australia
7	Ningaloo Vision	380	Australia
8	Cossak Pioneer	80	Australia
9	Umurao	120	New Zealand
10	Raroa	102	New Zealand

(Continued)

TABLE 4.5 (Continued)
Details of FPSO Commissioned Worldwide

S. No	Platform	Water Depth (m)	Location
North America			
1	Terra Nova	95	Canada
2	Sea Rose	122	Canada
3	Yuum K'ak'naab	100	US
Egypt			
1	Zaafarana	60	Egypt
2	PSVM	2000	Angola
3	Kizomba A	1241	Angola
4	Kizomba B	1163	Angola
5	Pazfor	762	Angola
6	CLOV	1365	Angola
7	Girassol	1350	Angola
8	Dalia	1500	Angola
9	Gamboa	700	Angola
10	Kuito	414	Angola
11	Petroleo Nautipa	137	Gabon
12	Knock Allan	50	Gabon
13	Abo	550	Nigeria
14	Bonga	1030	Nigeria
15	Armada Perkasa	13	Nigeria
16	Armada Perdana	350	Nigeria
17	Erha	1200	Nigeria
18	Usan	750	Nigeria
19	Agbami	1462	Nigeria
20	Akpo	1325	Nigeria
21	Ukpokiti	***	Nigeria
22	Kwame Nkrumah MV 21	***	Ghana
23	Sendje Ceiba	90	Equatorial Guinea
24	Aseng	945	Equatorial Guinea
25	Zafiro	***	Equatorial Guinea
26	Chinguetti Berge Helene	800	Mauritania
27	Baobab	1219	Cote d'Ivoire
Europe			
1	Huntington	91	UK
2	BW Athena	134	UK
3	Global producer III	140	UK
4	Bleo Holm	105	UK
5	Aoka Mizu	110	UK
6	Kizomba	1341	UK
7	Hummingbird	120	UK

(Continued)

TABLE 4.5 (Continued)
Details of FPSO Commissioned Worldwide

S. No	Platform	Water Depth (m)	Location
8	Petrojarl Foinaven	461	UK
9	Maersk Curlew	76	UK
10	Schiehallion	400	UK
11	North Sea Producer	125	UK
12	Caption	106	UK
13	Norne	380	Norway
14	Alvheim	130	Norway
15	Petrojarl I	100	Norway
16	Skarv	391	Norway
17	Goliat	400	Norway
18	Asgard A	300	Norway
19	Petrojarl Varg	84	Norway
South America			
1	Cidade de Rio das Ostras	977	Brazil
2	Cidade de Sao Mateus	763	Brazil
3	P-63	1200	Brazil
4	Frade	1128	Brazil
5	Cidade de Victoria	1400	Brazil
6	Peregrino	120	Brazil
7	Espadarte I	1100	Brazil
8	Espadarte II	850	Brazil
9	Golfinho	1400	Brazil
10	Marlim Sul (south)	1430	Brazil
11	Espirito Santo	1780	Brazil
12	Cidadde de Angra	2149	Brazil
13	Cidade de Niteroi MV18	1400	Brazil
14	Cidade de Santos MV20	1300	Brazil
Asia			
1	Bohai Shi Ji	20	China
2	Bohai Ming Zhu	31	China
3	Song Doc MV19	55	Vietnam
4	Ruby II	49	Vietnam
5	Ruby Princess	50	Vietnam
6	Arthit	80	Thailand
7	Bualuang	60	Thailand
8	Anoa Natuna	253	Indonesia
9	Kakap Natuna	88	Indonesia
10	Dhirubhai I	1200	India

4.7.3 SINGLE-POINT ANCHOR RESERVOIR (SPAR)

Spar platform belongs to the category of neutrally buoyant structures and consists of a deep draft floating caisson (Finn et al., 2003; Glanville et al., 1991; Graff, 1981a, 1981b). This caisson is a hollow cylindrical structure similar to a very large buoy (Ran et al., 1994; Rana and Soong, 1998; Zhang et al., 2007). Its four major components are hull, moorings, topsides, and risers. The spar relies on a traditional mooring system—i.e., anchor-spread mooring or catenaries mooring system, to maintain its position (Dawson, 1983; Rho et al., 2002, 2003). The spar design is now being used for drilling, production, or both. The distinguishing feature of a spar is its deep draft hull, which produces very favorable motion characteristics. The hull is constructed by using regular marine and shipyard fabrication methods, and the number of wells, surface wellhead spacing, and facility weight dictate the size of the center well and the diameter of the hull. Figure 4.12 shows a typical spar platform. In the classic or complete cylinder hull forms, the whole structure is divided into upper, middle, and lower sections. The upper section is compartmentalized around a flooded center well housing different types of risers; namely, production riser, drilling riser, and export/import riser. This upper section provides buoyancy for the spar. The middle section is also flooded but can be configured for oil storage. The bottom section, called the keel, is also compartmentalized to provide buoyancy during transport and to contain any field-installed, fixed ballast. The mooring lines are a combination of spiral strand wire and chain. The taut mooring system is possible due to the small motions of the spar and has a reduced scope, defined as the ratio of the length of the mooring line to water depth, and cost compared with a complete catenary system. Mooring lines are anchored to the seafloor with a driven or suction pile.

Different types of spars include (i) classical spar, which has a cylindrical hull with heavy ballast at the bottom of the cylinder, and (ii) truss spar, which has a shorter cylinder called a hard tank. The truss is further connected to its bottom to a soft tank, which houses ballast material, (iii) cell spar, which has a large central cylinder surrounded by smaller cylinders of alternating lengths. A soft tank is attached to the bottom of a longer cylinder to house ballast material.

The advantages of spar platforms are as follows: (i) low heave and pitch motion compared to other platforms; (ii) use of dry trees (i.e., on the surface); (iii) ease of fabrication; (iv) unconditional stability as its center of gravity is always lower than the center of buoyancy, resulting in a positive GM (metacentric height); and (v) derive no stability from its mooring system and hence does not list or capsize even when completely disconnected from its mooring system (Haslum and Faltinsen, 1999). A few disadvantages include the following: (i) installation is difficult as the hull and the topsides can only be combined offshore after the spar hull is upended; (ii) has little storage capacity, which brings along the necessity of a pipeline or an additional FSO; and (iii) have no drilling facilities. Figure 4.13 shows worldwide deepwater facilities.

4.8 NEW GENERATION OFFSHORE PLATFORMS

As the availability of oil and gas reserves moves toward higher water depths, oil and gas exploration is targeted at deep and ultra-deep waters. As the encountered

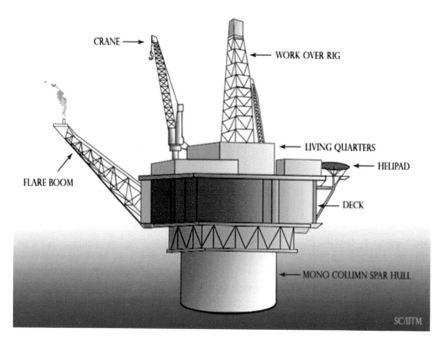

FIGURE 4.12   SPAR platform.

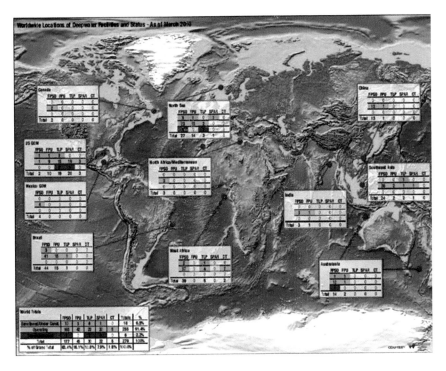

FIGURE 4.13   Deepwater facilities worldwide (Courtesy Offshore Magazine, 2016).

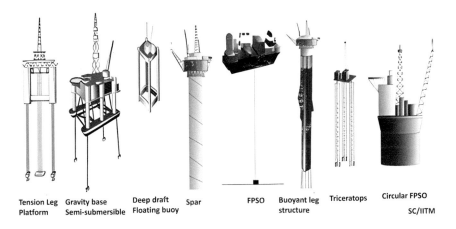

| Tension Leg Platform | Gravity base Semi-submersible | Deep draft Floating buoy | Spar | FPSO | Buoyant leg structure | Triceratops | Circular FPSO SC/IITM |

FIGURE 4.14   Types of ultra-deepwater structures.

environmental loads are more severe in greater water depths, the geometric form of offshore platforms proposed for deep and ultra-deep waters needs special attention (Gurley and Ahsan, 1998). Apart from being cost-effective, the proposed geometric form shall also have better motion characteristics under the encountered forces arising from the rough sea. Offshore structures that are found suitable for deep and ultra-deep waters are shown in Figure 4.14.

4.8.1 Buoyant Leg Structure (BLS)

Buoyant leg structures (BLSs) are tethered spars with single or group of cylindrical water-piercing hulls; these are alternative structural forms to TLPs and conventional spars (Capanoglu et al., 2002; Copple and Capanoglu, 1995; Robert and Capanoglu, 1995). They are positively buoyant, wherein the buoyancy exceeds the mass of the structure. Although being positively buoyant, positive metacentric height is maintained to ensure the desired structural stability, even after the removal of tethers from the structure. This characteristic ensures high stability and deep draft, which makes the structural form relatively insensitive to increased water depth. Since the BLS is a deep draft structure, the exposed structural part near the free surface is reduced, and the forces exerted on the structure are reduced when compared with the conventional TLPs (Chandrasekaran et al., 2015a). Since the risers are inside the moon pool of the BLS, the forces exerted on the risers are also minimized, but below the keel of the BLS, some forces like waves or currents act (Chakrabarti and Hanna, 1990; Chakrabarti, 1987, 1990, 1994, 2002, 1971, 1980). Halkyard et al. (1991) initially proposed the concept of a tension buoyant tower, which was subsequently modified by other researchers (Halkyard et al., 1991). The structural form of BLS is evolved by combining the advantageous features of spars and TLPs where its deep draft hull limits the vertical motion to a significant extent (Shaver et al., 2001); BLS resembles spar due to its shape and deep-draft feature, and its response behavior is similar to that of a TLP due to its restoring system (Gasim et al., 2008). BLS is simple to fabricate and easy to load out, tow, and install (Capanoglu et al., 2002; Sun et al., 1995). Figure 4.15

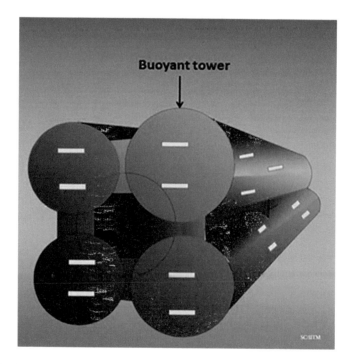

FIGURE 4.15   Buoyant tower in fabrication yard.

shows the views of the buoyant tower in the fabrication yard, while different stages of installation of BLS are shown in Figure 4.16.

The installation process of BLS is the combination of the installation procedures of the spar and TLP. Since spar is a stable structure, it is installed simply by free-floating, while TLP is generally installed by achieving required pretension in tethers using the following techniques: (i) ballast, (ii) pull-down, or (iii) both pull-down and ballast methods. During the installation of BLS, the structure can be free-floated using its permanent ballast. Pretension in the tethers can be achieved by the previously mentioned procedure. In the ballast method, the structure will be additionally ballasted until it achieves the required draft; tethers are then attached from the structure to the seafloor. Additional ballast will be removed from the structure to enable pretension in the tethers. In the pull-down method, the free-floating structure will be pulled down until it achieves the required draft; excess buoyancy that is transferred to the tethers helps to achieve the desired pretension. The pull-down and ballast method is a combination of the previously mentioned procedures. BLS imposes improved motion characteristics and more convenient riser systems, as they consist of simple hulls in comparison to spars or TLPs. BLS is more economical than TLPs or spars due to the reduced cost of commissioning. The first buoyant tower drilling production platform, CX-15, for Peru's Corvina offshore field was installed

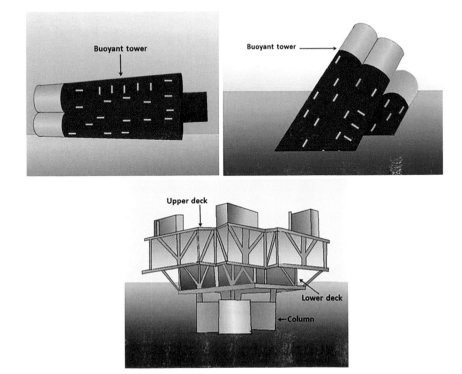

FIGURE 4.16 Load-out and installation of BLS.

in September 2012 at a water depth of more than 250 m with a production capacity of 12,200 barrels per day.

4.8.2 TRICERATOPS

More innovative geometric forms of offshore platforms have evolved in the recent past to improve the motion characteristics of these platforms under deep and ultra-deep waters. Triceratops, non-ship-shaped FPSOs, and Min Doc are a few of them. The conceptual idea of the triceratops discussed in the literature indicated favorable characteristics of the platform under deep and ultra-deep waters (Chandrasekaran et al., 2013; Chandrasekaran and Jamshed, 2015; Chandrasekaran et al., 2010; Chandrasekaran and Senger, 2016; Chandrasekaran et al., 2007a, 2011); Figure 4.17 shows the conceptual view of the triceratops. Geometric innovativeness imposed in the design by the introduction of ball joints between the deck and BLS makes triceratops different from other new-generation offshore platforms (Charles et al., 2005; Chaudhury and Dover, 1995). Triceratops consist of three BLS units, a deck, three ball joints between the BLS units and deck, and a restoring system either with a restraining leg or with the tethers (Chandrasekaran et al., 2015b). Ball joints transfer all translation but not rotation about any axis, making the platform different from other classic types of offshore structures; the distinct motion characteristics of its structural members, such as BLS and deck, provide uniqueness to its structural

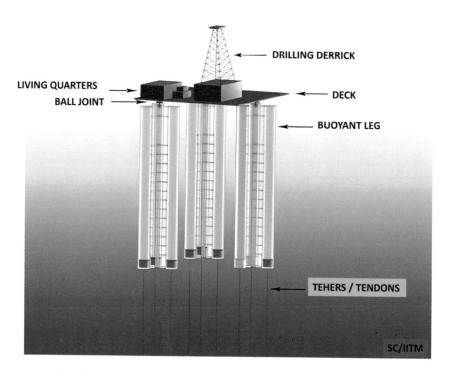

FIGURE 4.17   Triceratops: Conceptual view.

behavior under lateral loads. Common types of offshore platforms have rigid body motion due to the rigid connection between the members; this makes the platform respond as a single unit. As the triceratops is integrated with different structural elements, it behaves as a rigid body in all translations but not in rotations about any axis due to the presence of the ball joints; rotational responses of BLS differ from that of the deck.

Studies focusing on their response behavior become interesting as the responses of BLS and deck are to be dealt with separately, which is not a common practice in most offshore platforms. In addition, the derived geometric form has a few advantages: i) reduction in forces exerted on the platform due to the decrease in the exposed part of the structure near the free surface and (ii) protection of risers from lateral forces as they are located inside the moon pool of the BLS. The presence of ball joints between the deck and BLS restrains the transfer of the rotational motion of the deck from BLS; translation and rotation motion of BLS under the encountered environmental loads are significantly high. However, due to the deep draft of the BLS, there is more possibility of unusual corrosion. Corrosion challenges can be overcome by a few techniques, such as (i) frequent inspection using corrosion testing probes, (ii) use of sacrificial anodes, (iii) anticorrosive coatings, and (iv) use of cathodic protection. The salient advantages are (i) better motion characteristics; suitable for deep waters; (iii) improved dynamics in comparison to TLPs and spars; (iv) wells within the protected environment and are laterally supported; (v) simple structure; (vi) simple station keeping; (vii) easy to install and decommission (installation can be part by part or as a whole structure); (viii) reusable and relocated; (ix) simple restraining system (does not require high strength systems like TLPs); (x) highly stable structure; and (xi) relatively low cost.

4.8.3 Floating, Storage and Regasification Units (FSRUs)

Transportation of unprocessed crude from the drilling/exploratory platform to the onshore site involves expensive systems like transportation through pipes, large vessels, etc., which makes oil production more expensive. In particular, the offshore platforms located far offshore prove to be highly uneconomical. Critical components of FSRU consist of regasification equipment that transforms LNG at −160°C and high-pressure storage tanks, loading arms for receiving LNG, export manifolds, and seawater pumps that use seawater to re-gasify the LNG. FSRU is the more cost-effective alternative to meet the lower demand for LNG than traditional, land-based terminals. It contains a regasification unit, gas turbine with generator, air compressors, fuel pumps, fire water and foam systems, freshwater systems, cranes, lubrication oil system, lifeboats, and helipad.

The LNG is stored at −160°C in double-walled insulated tanks to limit boil-off. The outer walls of the tank are made of prestressed reinforced concrete or steel to limit the temperature during the storage period. Despite the high-quality insulation, a small amount of heat still penetrates the LNG tanks, causing minor evaporation. The resulting boil-off gas is captured and fed back into the LNG tank using compressor and recondensing systems. This recycling process prevents any natural gas from escaping the terminal under normal operating conditions. The LNG is subsequently extracted from the tanks, pressurized, and re-gasified using heat exchangers (Zhang and Dong, 2013). The tanks are equipped with submerged pumps that transfer the LNG toward other high-pressure pumps. The compressed LNG (at around 80 times atmospheric pressure) is then turned back into a gaseous state in vaporizers and once returned to its gaseous state, the natural gas is treated in several ways, including metering and odorizing before it is fed into the transmission network.

LNG is warmed using the heat from seawater. This is done in a heat exchanger (with no contact between the gas and the seawater), resulting in a slight drop in the temperature of the seawater, which reaches 6°C at the end of the discharge pipe, quickly becoming imperceptible once diluted. Natural gas is odorless. Although non-toxic, it is inflammable and is therefore odorized to ensure even the slightest leak can be identified. This is done by injecting tetrahydrothiophene (THT), which is an odorant detectable in very small doses, at the terminal before the natural gas is distributed. The gas turbine equipped at the topside of the FSRU uses multiple units of generating capacity of up to 10–12 MW. The instrument air system provides air for the plant and the instrument air in the process.

Inert gas (nitrogen) will be generated on demand by a membrane package using dry, compressed air. A backup inert gas supply system consisting of compressor seals, cooling medium, expansion drums, and utility stations also has to be provided. The oil pump provides high-pressure oil to the engine. The fuel is pumped from the fuel tank to the primary fuel filter/water separator, which is then pressurized to 650 kPa gauge pressure by the fuel transfer pump. The pressurized fuel is then sent through the secondary/tertiary fuel filter. Water supply for the fire-fighting systems is supplied by fire water pumps at a pumping rate of about 600–5000 m³/h at the discharge flange at a pressure of about 18 bar. A film-forming fluro protein (FFFP) concentrate system is provided to enhance the effectiveness of the deluge water spray that protects the separator module, which has a high potential for hydrocarbon pool fires. FFFP is a natural protein foaming agent that is biodegradable and non-toxic. The freshwater maker system will utilize a reverse osmosis process to desalinate the seawater at the rate of 5 m³/h. The saline effluent from the freshwater will be directed overboard through the seawater discharge caissons, while the freshwater will be stored in a freshwater tank. Water delivered to the accommodation module will be further sterilized in a UV sterilization plant before being stored in a potable water header tank. The lubrication system contains an oil cooler, oil filter, gear-driven oil pump, pre-lube pump, and an oil pan that meets offshore tilt requirements. The internal lubrication system is designed to provide a constant supply of filtered, high-pressure oil. This system meets the tilt requirements for nonemergency offshore operations. Lubrication oil should have unique features for offshore requirements, such as (i) water solubility; (ii) non-sheering on water surface; (iii) excellent lubrication properties; (iv) biodegradable; and (v) nontoxic to aquatic environment.

4.9 DRILLSHIPS

The drillship is an adaptation of a standard sea-going ship of mono-hull form; additions are a substructure containing a moon pool and cantilevers from which drilling operations may be carried out. Drill ships are designed to carry out drilling operations in deep sea conditions that are quite turbulent and susceptible to wave action. A typical drillship has a drilling platform and derrick located in the middle of its deck. Drill ships contain a hole called a moon pool, extending right through the ship, down the hull. They are relatively unstable and liable to be tossed by waves and currents. They use Dynamic Positioning System (DPS) and moorings for station keeping. The significant components of DPS include a controller, sensor system, thrusters, and power system. DPS comprises electric motors on the underside of the ship hull, and they are capable of propelling the ship in any direction. Propeller motors are integrated with the computer system of the ship. It uses the satellite positioning system technology, in conjunction with sensors located on the drilling template, to ensure that the ship is directly above the drill site at all times. DPS activates thrusters to move the ship back to its original position. The controllers command the action of the thrusters installed at the bottom of the ship hull and it keeps the platform within a tolerance radius of about 2–6% of the water depth. DPS controls displacement in the surge, sway, and yaw motion. In the mooring system, eight to ten anchor lines are required, and hence, uneconomical. Drill ships have the following advantages compared to semisubmersibles: (i) ship-shaped hull; (ii) subjected to longer periods of downtime under wind and wave action; (iii) used in smooth waters; and (iv) large load-carrying capacity.

4.10   ENVIRONMENTAL FORCES

Loads for which an offshore structure must be designed can be classified into the following categories:

- Permanent loads or dead loads.
- Operating loads or live loads.
- Other environmental loads, including earthquake loads.
- Construction and installation loads.
- Accidental loads.

While the design of buildings onshore is influenced mainly by the permanent and operating loads, the design of offshore structures is dominated by environmental loads, especially waves, and the loads arising in the various stages of construction and installation. In civil engineering, earthquakes are normally regarded as accidental loads. However, in offshore engineering, they are treated as environmental loads. Environmental loads are those caused by environmental phenomena that are random. These include wind, waves, currents, tides, earthquakes, temperature, ice, seabed movement, and marine growth. Their characteristic parameters, defining design load values, are determined in particular studies based on available data. According to U.S. and Norwegian regulations (or codes of practice), the mean recurrence interval for the corresponding design event must be 100 years, while according to British rules, it should be 50 years or greater. The different loads to be considered while designing the structure are wind loads, wave loads, mass, damping, ice loads, seismic loads, current loads, dead loads, live loads, impact loads, etc.

4.10.1   LOAD CLASSIFICATION

Loads for which an offshore structure must be designed can be classified into the following categories; namely, i) permanent loads or dead loads; ii) operating loads or live loads; iii) other environmental loads including earthquake loads; iv) construction and installation loads; and v) accidental loads. While the design of buildings onshore is usually influenced mainly by the permanent and operating loads, the design of offshore structures is dominated by environmental loads, especially waves, and the loads arising in the various stages of construction and installation. In civil engineering, earthquakes are normally regarded as accidental loads. But, in offshore engineering, they are treated as environmental loads. Environmental loads are those caused by environmental phenomena. These include wind, waves, currents, tides, earthquakes, temperature, ice, sea bed movement, and marine growth. Their characteristic parameters, defining design load values, are determined in special studies based on available data. According to U.S. and Norwegian regulations (or codes of practice), the mean recurrence interval for the corresponding design event must be 100 years, while according to British rules, it should be 50 years or greater.

4.10.2   WIND FORCES

Wind loads act on the portion of a platform above the water level as well as on any equipment, housing, derrick, etc., located on the deck. An important parameter of wind data is the time interval over which wind speeds are averaged. For averaging intervals of less than one minute, wind speeds are classified as gusts. For averaging intervals of one minute or longer, they are classified as sustained wind speeds. The wind velocity profile is given by [API-RP2A]:

$$V_h/V_H = \left(h/H\right)^{1/n} \tag{4.2}$$

where V_h is the wind velocity at height h, V_H is the wind velocity at reference height H, typically 10 m above mean water level and 1/n is 1/13 to 1/7, depending on the sea state, the distance from land, and the averaging time interval. It is approximately equal to 1/13 for gusts and 1/8 for sustained winds in the open ocean. From the design wind velocity V(m/s), the static wind force F_w(N) acting perpendicular to an exposed area A(m²) can be computed as follows:

$$F_w = \left(1/2\right)\rho_a V^2 C_s A \tag{4.3}$$

where ρ_a is the wind density (= 1.025 kg/m³), C_s is the shape coefficient. Typical values are 1.5 (for beams and sides of buildings), 0.5 for cylindrical sections, and 1.0 for the total projected area of the platform. Shielding and solidity effects can be accounted for in the judgment of the designer using appropriate coefficients. In cases where the member is inclined to the direction of the wind, the resulting force will act essentially in the direction normal to the member, and its magnitude will be given approximately by Equation (2.2) with the normal velocity component of the wind replacing the total wind velocity. Consider, for example, an inclined member shown in Figure 4.18.

If α denotes the angle between the wind direction and the normal to the member surface, the velocity component normal to the surface will be V cos, α and the equation for the force is given by:

$$F_w = (1/2)\,\rho_a\,C_s\,A\,(V\cos\alpha)^2 \tag{4.4}$$

where A denotes the area presented to the normal velocity component. For a circular cylinder of Length L and diameter D, or a flat plate of length L and width D, A=LD. For combination with wave loads, the DNV and DOE-OG rules recommend the most unfavorable of the following two loadings; namely, i) one-minute sustained wind speeds combined with extreme waves and ii) three-second gusts. API-RP2A distinguishes between global and local wind load effects. For the first case, it gives guideline values of mean one-hour average wind speeds to be combined with extreme waves and current. For the second case, it gives values of extreme wind speeds to be used without regard to waves. Wind loads are generally taken as static. When the ratio of height to the least horizontal dimension of the wind-exposed object (or structure) is greater than 5, then this object (or structure) could be wind-sensitive. API-RP2A requires the dynamic effects of the wind to be taken into account in this case, and the flow-induced cyclic wind loads due to vortex shedding must be investigated. The total wind load on the platform can be calculated using the wind blockage area and the pressure calculated as above. The shape coefficient (Cs) shall be selected as per AP RP2A guidelines (API-RP 2A, 2000, 2005).

But for the calculation of global wind load (for jacket and deck global analysis), the shape coefficient can be 1.0. The total force on the platform can be calculated as:

$$\begin{aligned} F_x &= f_w A_x C_s \\ F_y &= f_w A_y C_s \end{aligned} \tag{4.5}$$

In the foregoing, the exposed areas (Ax and Ay) shall be calculated as the product of (length and height) or (width and height) depending on the axis system followed. Figure 4.19 shows the wind load in oblique directions.

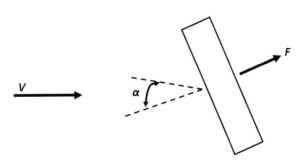

FIGURE 4.18   Wind force on an inclined member.

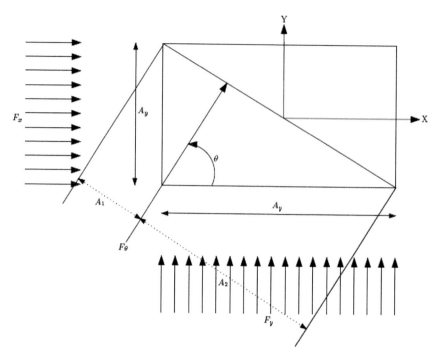

FIGURE 4.19 Wind load on oblique directions.

Wind load in oblique directions can be calculated using the following relationship:

$$F_\theta = F_x \cos\theta + F_y \sin\theta \qquad (4.6)$$

In practical design, it is often only F_x and F_y will be calculated and applied in the structural analysis as basic loads and the wind load effect due to non-orthogonal directions are simulated using factors in terms of Fx and F_y in the load combinations. The factors can be calculated as follows: The projected areas can be calculated as

$$A_1 = A_x \cos\theta$$
$$A_2 = A_y \sin\theta \qquad (4.7)$$

$$F_\theta = f_w (A_1 + A_2)$$
$$F_\theta = f_w (A_x \cos\theta + A_y \sin\theta)$$
$$F_{\theta x} = f_w [A_x \cos\theta + A_y \sin\theta] \cos\theta \qquad (4\,8)$$
$$F_{\theta y} = f_w [A_x \cos\theta + A_y \sin\theta] \sin\theta$$

where $F_{\theta x}$ and are the components of F_θ in x and y directions, respectively. The ratio between $F_{\theta x}$ and F_x can be expressed as:

$$\frac{F_{\theta x}}{F_x} = \frac{f_w \left[A_x \cos\theta + A_y \sin\theta \right] \cos\theta}{f_w A_x}$$

$$= \cos^2\theta + \left[\frac{A_y}{A_x}\right] \sin\theta \cos\theta \tag{4.9}$$

Similarly, the ratio between $F_{\theta y}$ and F_y can be expressed as:

$$\frac{F_{\theta y}}{F_y} = \frac{f_w \left(A_x \cos\theta + A_y \sin\theta \right) \cos\theta}{f_w A_y}$$

$$= \sin^2\theta + \left[\frac{A_x}{A_y}\right] \sin\theta \cos\theta \tag{4.10}$$

4.10.3 WAVE FORCES

Offshore structure: of all the environmental loadings, wave loading of an offshore structure is usually the most important one (Faltinsen, 1990; Faltinsen et al., 1995; Michel, 1999; Moan and Sigbjornson, 1977; Skjelbreia and Hendrickson, 1961; Soding et al., 1990; Stansberg et al., 2002, 2004; Fugazza and Natale, 1992; Sellers and Niedzwecki, 1992). Forces on offshore structures are caused by the motion of water particles due to waves (Bea et al., 1999; Bearman, 1984; Moe and Verley, 1980; Mogridge and Jamieson, 1975). Determination of these forces requires the solution of two separate problems; namely, i) firstly, computation of wave kinematics by an appropriate wave theory; and ii) secondly, computation of wave forces on individual members and the total structure from the computed wave kinematics (Blevins, 1994; Boaghe et al., 1998; Brika and Laneville, 1993; Chandrasekaran, 2013a, 2013c, 2014a; Pilatto, 2003; Pilatto and Stocker, 2002, 2003). Two different analysis concepts are used; namely, a) single design wave analysis and b) random wave analysis (Chakrabarti, 1984, 1998, 2005; Chakrabarti et al., 1976). In a single design wave, a design wave in terms of a regular wave of a given height and period is defined, and the forces due to this wave are calculated using a high-order wave theory (Chakrabarti and Tam, 1975; Chandrasekaran, 2015b, 2015d). Usually, the 100-year wave—i.e., the maximum wave with a return period of 100 years—is chosen. No dynamic analysis of the structure is carried out. Static analysis is appropriate when the dominant wave periods are well above the period of the structure and under extreme storm waves acting on shallow water structures. In random wave analysis, statistical analysis is carried out based on a wave scatter diagram for the location of the structure (Bringham, 1974). Appropriate wave spectra are defined to perform the dynamic analysis in the frequency domain. With statistical methods, the most probable maximum force during the lifetime of the structure is calculated using linear wave theory (Moharammi and Tootkaboni, 2014; Montasir and Kurian, 2011).

In the calculation of wave forces, there are two different domains of problems encountered. One domain pertains to wave forces on small bodies wherein the incoming waves are not modified by the presence of the structure, and wave forces are estimated using a semi-empirical approach (Ertas and Eskwaro-Osire, 1991;

Hogben and Standing, 1974; Hove and Foss, 1974; Hsu, 1981). The other domain
pertains to wave forces on larger bodies wherein the incoming waves are modified
by the presence of the structure, and wave forces are estimated by diffraction theory
(Isaacson, 1978b; Isaacson et al., 1982, 1998, 2000).

4.11   WAVE THEORIES

Wave theories based on certain assumptions like fluid ideal and incompressible, flow
irrotational, constant atmospheric pressure, etc., describe the kinematics of waves of
water based on potential theory. By solving a boundary value problem, a solution is
sought for velocity potential, which is a scalar function of space and time whose gradient yields the velocity vector. Wave theories serve to calculate the particle velocities and accelerations and the dynamic pressure as functions of the surface elevation
of the waves. For long-crested regular waves, the flow can be considered two-dimensional and is characterized by the parameters: wave height (H), period (T), and water
depth (d), as shown in Figure 4.20. $k = 2\pi / L$ denotes the wave number, $\omega = 2\pi / T$
denotes the wave circular frequency, and f denotes cyclic frequency.

Different wave theories of varying complexity, developed based on simplify-
ing assumptions, are appropriate for different ranges of wave parameters. Among
the most common theories are namely: i) Linear or first order or Airy theory;
ii) Stokes fifth-order theory; iii) Solitary wave theory; iv) Cnoidal theory; v) Dean's
stream function theory; and vi) Numerical theory by Chappelear (Stokes, 1880;
Spanos and Agarwal, 1984). Figure 4.21 shows the chart for the selection of the
most appropriate theory based on the parameters H, T, and d. For example, Linear

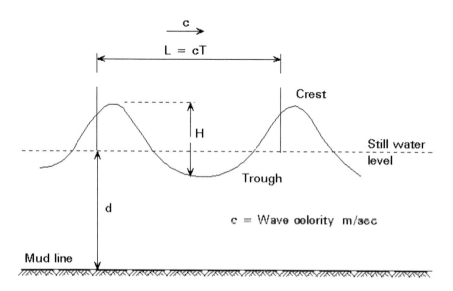

FIGURE 4.20   Wave parameters.

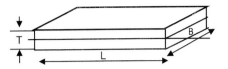

FIGURE 4.21   Displacement of a prismatic structure.

wave theory can be applied when $H/gT^2 < 0.01$ and $d/GT^2 > 0.05$, besides other ranges as described in the literature.

4.11.1   AIRY'S THEORY

Airy's theory is also known as linear or first-order or small-amplitude or regular-wave theory. Regular waves begin with the simplest mathematical representation assuming ocean waves are two-dimensional (2-D), small in amplitude, sinusoidal, and progressively definable by their wave height and period in a given water depth. In this simplest representation of ocean waves, wave motions, and displacements, kinematics (that is, wave velocities and accelerations), and dynamics (that is, wave pressures and resulting forces and moments) will be determined for engineering design estimates. When wave height becomes larger, the simple treatment may not be adequate. The next part of the regular waves section considers a 2-D approximation of the ocean surface to deviate from a pure sinusoid. This representation requires using more mathematically complicated theories. These theories become nonlinear and allow the formulation of waves that are not purely sinusoidal in shape; for example, waves having flatter troughs and peaked crests are typically seen in shallow coastal waters when waves are relatively high. Gravity waves may also be classified by the water depth in which they travel. The following classifications are made according to the magnitude of relative depth, d/L. The relative depth is the ratio of water depth to wavelength. Water waves are classified in Table 4.6 based on the relative depth criterion d/L as deep water waves, transitional water waves, and shallow water waves. The resulting limiting values taken by the function tanh (kd) are: i) the argument of the hyperbolic tangent $(kd) = 2\pi d/L$ gets large and tanh(kd) approaches 1; and ii) for small values of kd, tanh (kd) is approximately equal to kd.

4.12   WATER PARTICLE KINEMATICS

Ocean surface waves refer generally to the moving succession of regular humps and hollows on the ocean surface. They are preliminarily generated at any offshore site by the drag of wind on the water's surface. For engineering purposes, it is customary to analyze the effects of surface waves on the structures, either using a single design wave chosen to represent the extreme storm conditions in the area of interest or by use of a statistical representation of the waves during extreme storm conditions. As discussed by Dawson (1983), it is necessary to relate the surface wave data to water velocity, acceleration, and pressure beneath the waves, which can be achieved by using the appropriate wave theory. Airy's theory presents a relatively simple theory of

TABLE 4.6

Classification of Water Waves

Classification	d/L	kd	tanh(kd)
Deepwater	½ to ∞	π to ∞	~ 1
Transitional	1/20 to ½	π/10 to π	tanh(kd)
Shallow water	0 to 1/20	0 to π/10	~ kd

wave motion, assuming a sinusoidal waveform. The wave amplitude (H) is assumed to be smaller than the wavelength (L) and water depth (d).

Although not strictly applicable to typical design waves used in offshore structural engineering, the Airy theory is valuable for preliminary calculations and for revealing the basic characteristics of wave-induced water motion. It also serves as a basis for the statistical representation of waves and induced water motion during storm conditions. According to Airy's wave theory, the sea surface elevation η (x, t) is given by:

$$\eta(x,t) = \frac{H}{2}\cos(kx - \omega t) \tag{4.11}$$

where, $\omega = \dfrac{2\pi}{T}$

The horizontal velocity $\dot{u}(x,t)$ and vertical velocity $\dot{v}(x,t)$ of the water particle at any location (x, y) and time t are given by Eqs. (4.12) and (4.13), respectively.

$$\dot{u}(x,t) = \frac{\omega H}{2}\frac{\cosh ky}{\sinh kd}\cos(kx - \omega t) \tag{4.12}$$

$$\dot{v}(x,t) = \frac{\omega H}{2}\frac{\sinh kY}{\sinh kd}\sin(kx - \omega t) \tag{4.13}$$

For small wave heights considered by the Airy theory, the horizontal water particle acceleration $\ddot{u}(x,t)$ and vertical water particle acceleration $\ddot{v}(x,t)$ at any location (x, y) and time t are given by:

$$\ddot{u}(x,t) = \frac{\omega^2 H}{2}\frac{\cosh kY}{\sinh kd}\sin(kx - \omega t) \tag{4.14}$$

$$\ddot{v}(x,t) = -\frac{\omega^2 H}{2}\frac{\sinh kY}{\sinh kd}\cos(kx - \omega t) \tag{4.15}$$

The free surface boundary conditions are linearized with the description of Airy's theory. The ocean bottom is considered flat and the kinematics are defined in the infinite two-dimensional field bounded by the bottom and still water level (SWL), as described by Chakrabarti (1990). The inherent assumption in the derivation of linear theory has a limit of $Y = d$ and therefore does not allow the computation above the SWL. Therefore, to extend the computations to the changing surface effect due to variable submergence, stretching modifications were suggested by Chakrabarti (1971). The horizontal wave particle velocity and acceleration are given by the following expression:

$$\dot{u}(x,t) = \frac{\omega H}{2} \frac{\cosh(kY)}{\sinh(k(d+\eta))} \cos(kx - \omega t) \tag{4.16}$$

$$\ddot{u}(x,t) = \frac{\omega^2 H}{2} \frac{\cosh(kY)}{\sinh(k(d+\eta))} \sin(kx - \omega t) \tag{4.17}$$

The velocity potential does not satisfy the Laplace equation, but it satisfies the dynamic free surface boundary condition. Therefore, in many physical situations, the linear theory even with stretching modifications is not adequate to describe water particle kinematics completely. Hence, same higher-order theories are required to obtain better free surface and water particle kinematics expressions.

4.13   HYDROSTATIC STABILITY OF FLOATING BODIES

All offshore structures have to be checked for static stability. The static equilibrium of a floating vessel is influenced by weight and buoyancy forces. The center of gravity is the weight (mass) center of the body about which the weight (mass) distribution is balanced (zero weight moment). Weight is the product of mass and gravitational acceleration (Winterstein, 1988; Chandrasekaran et al., 2007, 2010; Chandrasekaran and Madhuri, 2015); it acts downwards through the center of gravity. Buoyancy is the weight of the displaced volume of water by the body generally at its equilibrium position; it acts upwards through the center of gravity. When a vessel is floating freely, these two forces must act along the same vertical line and counteract each other. Stability is defined as the ability of a system to return to its undisturbed position after external force is removed. The higher the value of the righting capacity (moment), the higher the stability of the vessel. Consider the following examples: a) Figure 4.21 shows the displacement of a prismatic structure of dimension (LxB) whose depth of immersion in T. Displaced volume is given by (LBT). Figure 4.22 shows the displacement of a ship. Displacement of a ship-shaped vessel is difficult

FIGURE 4.22   Displacement of a ship.

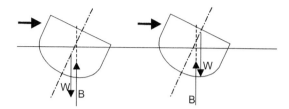

Case 1: Positively stable   Case 2: Negatively stable or unstable

FIGURE 4.23 Cases of stability.

to compute as the ship is contoured. Usually, it is obtained by rigorous calculation from the ship contour charts. However, for additional computational purposes, each ship type is represented by an equivalent block coefficient with a prismatic box based on the shape and given as (LBTC$_B$), where C_B is the block coefficient of the vessel.

4.13.1 Transverse Stability

Transverse stability is determined by the points of action of weight (the center of gravity) and buoyancy (the center of buoyancy) and the horizontal distance and relative position between these two parameters. Examine the two cases in Figure 4.23: Case 1 is stable as net moment tends to right the body which is called positively stable; Case 2 is negatively stable or unstable as net moment tends to destabilize the body. Figure 4.24 shows the definition sketch for a righting moment of a vessel and metacenter. In this figure, K is the keel (bottom point/line) of the vessel, where G is the point of action of weight—i.e. center of gravity—and B is the point of action of buoyancy—i.e. center of buoyancy. Let us assume that the vessel heels by an angle θ, shifting the center of buoyancy from B to B1. At this orientation, a couple acting on the vessel is given (M = WxGZ). Metacenter (M) is the point of intersection between the line of action of buoyancy force (vertical) and the centerline of the vessel in its inclined position; this will be likely the center of oscillation of a suspended pendulum. In such a situation, GM becomes the length of the string. For the pendulum to swing in a stable oscillation and return to its original position, the center must be located above the pendulum. The metacentric height is the sum of the distance of the vessel keel to the center of buoyancy (KG) and the distance between the center of buoyancy and the metacenter (BM). The moment now becomes (W x GM sinθ). When M is above G, the moment is righting and is stable. If it is below G, then this shall result in overturning and becomes unstable. For inclination lesser than 15°, BM will be the second moment of area (Moment of inertia) of the water plane cross-sectional area about the middle line and is given by:

$$BM = \frac{I_{xx}}{\nabla} \qquad (4.18)$$

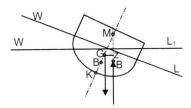

FIGURE 4.24   Righting moment of a vessel and metacenter.

For GM greater than zero, the floating system shall be positively stable. Therefore, for a submerged object to be stable, the center of gravity must be below the center of buoyancy. Since the point of action of buoyancy is fixed along the line of gravity and does not change, the metacenter is B itself, and therefore, the condition that GM is greater than zero still holds good.

4.14   BUOYANT FORCES

Pressure loading on fully or partially submerged objects arises from the weight of the water above it and the movement of the water around it by wave action. The subsurface pressure gives the magnitude of pressure as determined by Airy's theory. One effect of pressure on a submerged member of an offshore structure is to induce stresses in the member. Another effect is to exert horizontal and vertical forces on the member. Forces arising from pressure associated with wave action are included in the Morison equation. However, an additional buoyant force also arises from the hydrostatic pressure and is given by:

$$p = -\rho g z \tag{4.19}$$

where ρ denotes the specific weight of the water and z denotes the vertical distance from the still water level which is negative. This force exists even when wave action is absent and must be accounted for separately. Offshore structural members are mostly made buoyant by air-tight sealing of the welds to avoid water entry. This is purposely planned so that the overall structure will have adequate buoyancy during installation. A typical example is the jacket structure. This kind of structure requires at least a reserve buoyancy of 10% to 15%. The reserve buoyancy is defined as buoyancy over its weight. To obtain this buoyancy, structural tubular members are carefully selected such that the ratio of buoyancy to weight is greater than unity. This means that the member will float in water. On the other hand, if the member is part of a structure supported at its two ends and forced to be submerged by the weight of other members, this member will experience an upward force equal to the displaced volume of water. This is called the buoyancy force. The buoyancy force can be calculated by two methods; namely, i) the marine method and ii) the rational method. The marine method assumes that the member is considered to have rigid body motion; it means that the weight of the member is calculated using submerged density of steel and applied to the member vertically down as a uniformly distributed load. This buoyant weight per unit length is given by:

$$W_B = \frac{1}{4}\pi\left(D^2 - (D-t)^2\right)(\rho_{steel} - 1.025) \tag{4.20}$$

where ρ_s is the density of steel. Unlike gravity force which is true body force acting on every particle of the body, buoyancy is the resultant of fluid pressure acting on the surface of the body. These pressures can only act normally on the surface. The rational method takes into account this pressure distribution on the structure that results in a system of loads consisting of distributed loads along the members and concentrated loads at the joints. The loads on the members are normal to the member axis and in the vertical plane containing the member. The magnitude of this distributed member load can be expressed as:

$$B_B = \frac{1}{4}\pi D^2 \rho_w \cos\alpha \tag{4.21}$$

where α is the angle between the member and its projection on a horizontal plane. Joint loads consist of forces acting in the direction of all the members meeting at a joint. These joint forces act in a direction that would compress the corresponding members if they acted directly on them and have magnitude as given as follows:

$$P_B = \rho_w Ah \tag{4.22}$$

where A is the displaced area which is equal to the area of the flooded members and h is the water depth at the end of the member under consideration.

4.15   CURRENT FORCES

Currents are caused by tides and storms besides circulations. In the absence of reliable field measurements, current velocities may be obtained from various sources. Current affects the wave period and its kinematics (Niedzwecki et al., 2000; Morison, 1953). In offshore platform design, the effects of current superimposed on waves are taken into account by adding the corresponding fluid velocities vectorially. Since the drag force varies with the square of velocity, this addition can greatly increase the forces on the platform. For slender members, cyclic loads induced by vortex shedding may also be important and should be examined (Khalak and Williamson, 1991; Humphries and Walker, 1987; Govardhan and Williamson, 2000). This force is computed using the drag component of the Morison equation with appropriate modifications of fluid kinematics.

4.16   ADDITIONAL ENVIRONMENTAL LOADS

4.16.1   EARTHQUAKE LOADS

Offshore structures in seismic regions are typically designed for two levels of earthquake intensity; namely, the strength level and the ductility level earthquake. A strength level earthquake is defined as having a "reasonable likelihood of not being exceeded during the platform's life" (mean recurrence interval ~ 200 - 500 years), and the structure is designed to respond elastically. A ductility level earthquake is defined as close to the "maximum credible earthquake" at the site and the structure is designed for inelastic response and to have adequate reserve strength to avoid collapse. For strength level design, the seismic loading may be specified either by sets of accelerograms or using design response spectra. If the design spectral intensity, characteristic of the seismic hazard at the site, is denoted by a_{max}, then API-RP2A recommends using a_{max} for the two principal horizontal directions and $0.5a_{max}$ for the vertical direction. The DNV rules, on the other hand, recommend a_{max} and $0.7\,a_{max}$ for the two horizontal directions (two different combinations) and $0.5\,a_{max}$ for the vertical. The value of a max and often the spectral shapes are determined by site-specific seismological studies. Designs for ductility-level earthquakes will normally require inelastic analyses for which the seismic input must be specified by sets of 3-component accelerograms, real or artificial, representative of the extreme ground motions that could shake the platform site. The characteristics of such motions, however, may still be prescribed using design spectra, which are usually the result of a site-specific seismo-tectonic study.

4.16.2   ICE AND SNOW LOADS

Ice and snow loads are not a concern in the tropical and equatorial regions. Ice is a primary problem for marine structures in the arctic and sub-arctic zones. Ice formation and expansion can generate large pressures that give rise to horizontal as well as vertical forces. In addition, large blocks of ice driven by current, winds, and waves with speeds that can approach 0.5 to 1.0 m/s may hit the structure and produce impact loads. Statically applied horizontal ice forces may be estimated as approximately follows:

$$F_i = C_i f_c A \qquad (4.23)$$

where A is the exposed area of the structure, f_c is the compressive strength of ice and C_i is the coefficient accounting for shape, rate of load application, and other factors with usual values between 0.3 and 0.7. Besides, ice formation and snow accumulations increase gravity and wind loads; the latter, by increasing areas exposed to the action of wind. Eurocode 1 gives more detailed information on snow loads.

4.16.3   TEMPERATURE VARIATIONS

Temperature variations in the regions of the location of offshore structures cause thermal stresses in these structures. To estimate these stresses, extreme values of sea and air temperatures that are likely to occur during the life of the structure must be estimated. Relevant data for the North Sea are given in BS 6235. Thermal stresses can also be induced by human factors—e.g., through accidental release of cryogenic material—which must be taken into account in design as accidental loads. The temperature of the oil and gas produced must also be considered.

4.16.4 MARINE GROWTH

Marine growth, or biofouling, is the ubiquitous attachment of soft and hard particles on the surface of a submerged structure. These range from seaweeds to hard-shelled barnacles. Its growth on the surface of the structure increases its diameter and affects its roughness. Its main effect is to increase the wave forces on the members by increasing not only exposed areas and volumes, but also, the drag coefficient due to higher surface roughness. In addition, it increases the unit mass of the member, resulting in higher gravity loads and lower member frequencies. Depending upon geographic location, the thickness of marine growth can reach 0.3 m or more. It is accounted for in design through appropriate increases in the diameters and masses of the submerged members.

4.16.5 TIDES

Tides are caused by to gravitational attraction of the sun and moon on Earth and it affects coastal areas by submergence during the high tide period. The tides are classified as (a) astronomical tides and (b) storm surges. Astronomical tides are caused essentially by the gravitational pull of the moon and the sun and storm surges are caused by the combined action of wind and barometric pressure differentials during a storm (Dean and Dalrymple, 2000). The combined effect of the two types of tide is called the storm tide. Tide-dependent water levels and the associated definitions, as used in platform design, are shown in Figure 4.25. The astronomical tide range depends on the geographic location and the phase of the moon. Its maximum, the spring tide, occurs at the new moon. The range varies from centimeters to several meters and may be obtained from special maps. Storm surges depend upon the return period considered and their range is on the order of 1 to 3m. When designing a platform, extreme storm waves are superimposed on the still water level (see Figure 4.25) while for design considerations such as levels for boat landing places, barge fenders, upper limits of marine growth, etc., daily variations of the astronomical tide are used.

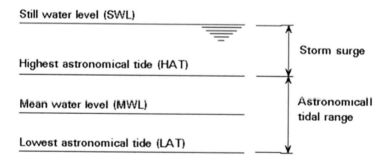

FIGURE 4.25 Tide-related definitions of sea surface level.

4.16.6 SEAFLOOR MOVEMENT

As a result of active geologic processes, storm wave pressures, earthquakes, pressure reduction in the producing reservoir, etc., sea floor movements occur. Loads generated by such movements affect, not only the design of the piles but the jacket as well. Such forces are determined by special geotechnical studies and investigations.

4.17 WIND FORCE ON COMPLIANT PLATFORM

Wind forces are attracted by the superstructure of the offshore deck. Wind forces are quantified by dividing the projected area of the superstructure into several segmental areas and velocity fluctuations are defined at the centroid of these areas. Wind fluctuations depend on space and time but the spatial fluctuations are normally neglected for compliant type structures. It may be safer to regard wind velocity as uniform along the face line normal to the wind direction. Mean and fluctuating wind forces are estimated in the surge direction which is instantly the wave heading direction as well. Direct wind pressure on the superstructure causes translational surge force and moments in the yaw and pitch directions. Due to the coupled nature of heave degree-of-freedom with surge, yaw, and pitch, heave motion is also activated. This coupled wind force also influences tether tension variation. As it is assumed that the total wind force is said to be concentrated on the aerodynamic center (AC) while the mass is lumped at the center of gravity (CG), the difference between the AC and CG activates the pitch and yaw forces in the deck. The wind-induced force given as drag per unit area, projected on a plane, normal to the mean wind velocity is given by:

$$f(y,z,t) = \frac{1}{2}\rho_a C_a(y,z) A_a \left[\dot{u}(y,z,t) - \dot{x}(t) \right]^2 \qquad (4.24)$$

where, $f(y,z,t)$ is the force per unit area and is a function of space (y,z) and time(t), ρ_a is the air density, A_a is the exposed area in surge direction, $C_a(y,z)$ is the force coefficient at elevation z and horizontal co-ordinate y, $\dot{u}(y,z,t)$ is the wind velocity in the surge direction varying with time, $\dot{x}(t)$ is the structural velocity in the surge direction. It is assumed that the directions of wind and surge motion are collinear. The wind velocity is expressed as:

$$\dot{u}(y,z,t) = \dot{u} + \dot{u}'(y,z,t) \qquad (4.25)$$

where, $\dot{u}$ is the mean wind velocity at 10m above MSL and $\dot{u}'(y, z, t)$ is the fluctuating wind velocity. Hence the force due to wind in N segments above the path segment (i.e.) starting from MSL, shall be expressed as:

$$F(y,z,t) = \sum_{p}^{N} \overline{C}_p \left[\dot{u} + \dot{u}'(y,z,t) - \dot{x}(t) \right]^2 \qquad (4.26)$$

$$\overline{C}_p = \frac{1}{2}\rho_A C_p A_{ap} \qquad (4.27)$$

where A_{ap} is the projected area of the pth segment, C_p is the aerodynamic force coefficient for the pth segment and N is the total number of segments. The mean wind velocity using the logarithmic law is as follows:

$$\bar{u}(z) = u(z_{ref}) \left\{ \frac{\ln\left[\dfrac{z}{z_0}\right]}{\ln\left[\dfrac{z_{ref}}{z_0}\right]} \right\} \qquad (4.28)$$

where z_{ref} is the reference elevation usually taken as 10m above mean sea level, z_0 is the roughness length and $\bar{u}(z)$ is the mean velocity at elevation of the vertical coordinate z. Specifying the value of C_p with roughness length, z_0 over the sea surface as:

$$C_p = \left[\frac{K}{\ln\dfrac{10}{z_0}} \right]^2 \qquad (4.29)$$

where K is the Von Karman Constant (k = 0.4). The above equation gives a conservative estimate of the mean wind velocity component due to the presence of water waves as compared to flow above a rigid surface. Fluctuating wind is estimated based on single point simulation; random variation of wind velocity fluctuation is assumed to be constant all through the projected area. This fluctuating velocity component is estimated by the Fourier synthesis of the wind spectrum suited to the offshore environment. Researchers (Kareem and Datton, 1982; Davenport, 1961a, 1961b) proposed wind spectra to describe the longitudinal wind velocity fluctuation. As the deep water-compliant type platforms have the frequencies of interest at very low values, the ordinates of this spectrum are maximum (Buchner and Bunnik, 2007; Buchner et al., 1999; Burrows et al., 1992). The shape of the spectrum in the very low-frequency range has very little effect on the design of land-based structures of fixed type, while it significantly influences the design of triceratops (Chandrasekaran and Nannaware, 2013). Expression for the fluctuating wind velocity spectrum is given by:

$$\left\{ \frac{nS_u(z,n)}{u_*^2} \right\} = a_1 f + b_1 f^2 + d_1 f^3 \qquad (0 < f \leq f_m)$$

$$= c_2 + a_2 f + b_2 f^2 \qquad (f_m < f < f_s) \qquad (4.30)$$

$$= 0.26 f^{-2/3} \qquad (f \geq f_s)$$

where,

$$u_* = \frac{K u(10)}{\ln\dfrac{10}{z_0}} \qquad (4.31)$$

$$a_1 = \frac{4L_u\beta}{z} \tag{4.32}$$

$$\beta_1 = 0.26\,f_s^{-2/3} \tag{4.33}$$

$$f = \frac{nz}{u(z)} \tag{4.34}$$

where f is a non-dimensional frequency

$$b_2 = \frac{\frac{1}{3}a_1f_m + \left[\frac{7}{3} + \ln\frac{f_s}{f_m}\right]\beta_1 - \beta}{\frac{5}{6}\left(f_m - f_s\right)^2 + \frac{1}{2}\left(f_m^2 - f_s^2\right) + 2f_m\left(f_s - f_m\right) + f_s\left(f_s - 2f_m\right)\ln\frac{f_s}{f_m}} \tag{4.35}$$

$$a_2 = -2b_2 f_m \tag{4.36}$$

$$d_1 = \frac{2}{f_m^3}\left[\frac{a_1 f_m}{2} - \beta_1 + b_2\left(f_m - f_s\right)^2\right] \tag{4.37}$$

$$b_1 = -\frac{a_1}{2f_m} - 1.5 f_m d_1 \tag{4.38}$$

$$c_2 = \beta_1 - a_2 f - b_2 f_s^2 \tag{4.39}$$

In the foregoing expressions, Z_0 is generally taken as 0.001266m and L_u is taken as 180m. For offshore decks, u (10) values are generally taken as 35 m/s, 40 m/s, and 45 m/s, and the corresponding values of u_* are 1.56m/s, 1.76m/s, and 2.01m/s. The wind velocity spectrum so produced covers the largely varying frequencies of compliant offshore platforms. For ease of simulation, the cross-correlation coefficient is assumed unity—i.e., the wind velocities are assumed to be fully correlated along the height of the superstructure. Thus, the time histories of wind velocities at different can be generated with the help of only the wind velocity spectrum defined above. Fluctuating wind represented by Emil Simiu's spectrum is simulated using the Monte–Carlo procedure.

4.18 DEAD LOAD

Dead load is the weight of the overall platform in the air, which includes piling, superstructure, jacket, stiffeners, piping, conductors, corrosion anodes, deck, railing, grout, and other appurtenances. Dead load excludes the following: weight of the drilling equipment placed on the platform including the derrick, draw works, mud pumps, mud tanks, etc.; the weight of production or treatment equipment located on the platform, including separators, compressors, piping manifolds, and storage tanks; the weight of drilling supplies that cause variable loads during drilling such as drilling mud, water, fuel, casing, etc.; weight of treatment supplies employed during production such as fluid in the separator, storage in the tanks;

drilling load, which is approximate combination of derrick load, pipe storage, rotary table load, etc.

4.19 LIVE LOAD

Live loads are acting in addition to the equipment loads. They include load caused by impacts of vessels and boats on the platform. Dynamic amplification factor is applied to such loads to compute the enhanced live loads. Live loads are generally designated as factor times of the applied static load. These factors are assigned by the designer depending on the type of platform. Table 4.7 gives the live load factors that are used in the platform design.

4.20 IMPACT LOAD

For structural components that experience impact under live loads, the stipulated live loads in Table 4.7 should be increased by an impact factor, as given in Table 4.8. Deck floor loads can be taken as 11.95 kN/m^2 in the drilling rig area, 71.85 kN/m^2 in the derrick area, and 47.9 kN/m^2 for pipe racks, power plants, and living and accommodation areas.

4.21 GENERAL DESIGN REQUIREMENTS

The design methodology of offshore platforms differs with different types of offshore structures. For example, vertical deformation will be lesser in the case of bottom-supported structures like jacket platforms, GBS, etc. Such platforms are highly rigid and tend to attract more forces. Hence, the design criteria should be to limit the stresses in the members. Displacement of the members under the applied loads will be insignificant. On the contrary, compliant structures are more flexible, as they all displaced more under wave action. They also create more disturbances in the waves. Hence, the design criteria will be to control displacement instead of limiting the stresses in the members. Orientation of the platform is another important aspect of the design. Preferred orientation is that members are oriented to have a less projected area to the encountered wave direction. This induces a lesser response from the members. The predominant wave direction for the chosen site is made available

TABLE 4.7
Typical Live Load Values Used in Platform Design (Graff, 1981a, 1981b)

Description	Uniform Load on Decks (kN/m²)	Concentrated Load on Deck	Concentrated Load on Beams
Walkway, stair	4.79	4.38 kN/m^2	4.45 kN/m^2
Areas > 40 m^2	3.11	–	–
Areas for light use	11.9	10.95 kN/m^2	267 kN

TABLE 4.8
Impact Factor for Live Loads

Structural Item	Load Direction	
	Horizontal	Vertical
Rated load in craned	20%	100%
Drilling hook loads	–	–
Supports of light machinery	–	20%
Supports of rotating machinery	50%	50%
Boat landings (kN)	890	890

to the designer based on which the platform orientation is decided (Chandrasekaran and Bhattacharyya, 2011). Following is the list of data required for the design of offshore structures:

- Land topographical survey of sufficient area covering the chosen site for platform installation.
- Hydrographical survey of the proposed location (hydrographic charts are used for this purpose).
- Information regarding silting at the site.
- Wind rose diagram showing information on wind velocities, duration, pre-pre-dominant direction around the year.
- Cyclonic tracking data showing details of the past cyclonic storm such that wind velocities, direction, peak velocity period, etc., are indicated.
- Oceanographic data including general tide data, tide table, wave data, local current, seabed characteristics, temperature, rainfall, and humidity.
- Seismicity level and values of acceleration.
- Structural data of existing similar structures, preferably in the close vicinity.

4.21.1 SOIL INVESTIGATION

The analysis of an offshore structure is an extensive task, embracing consideration of the different stages—i.e., execution, installation, and in-service stages, during its life. Many disciplines such as structural, geotechnical, naval architecture, and metallurgy are involved. The analytical models used in offshore engineering are in some respects similar to those adopted for other types of steel structures. Only the salient features of offshore models are presented here. The same model is used throughout the analysis with only minor adjustments to suit the specific conditions—e.g., at supports, in particular, relating to each analysis. Stick models (beam elements assembled in frames) are used extensively for tubular structures (jackets, bridges, flare booms) and lattice trusses (modules, decks). Each member is (normally) rigidly fixed at its ends to other elements in the model. If more accuracy is required, particularly for the assessment of natural vibration modes, the local flexibility of the connections may be represented by a joint stiffness matrix. In addition to its geometrical and material properties, each member is characterized by hydrodynamic coefficients—e.g., relating to drag, inertia, and marine growth—to allow wave forces to be automatically generated (Kim et al., 1997; Kim and Zou, 1995). Integrated decks and hulls of floating platforms, involving large bulkheads, are described by plate elements. The characteristics assumed for the plate elements depend on the principal state of stress to which they are subjected. Membrane stresses are taken when the element is subjected merely to axial load and shear. Plate stresses are adopted when bending and lateral pressure are to be taken into account. After developing a preliminary model for analysis, member stresses are checked for preliminary sizing under different environmental loads.

Verification of an element consists of comparing its characteristic resistance(s) to a design force or stress. It includes (i) a strength check where the characteristic resistance is related to the yield strength of the element and (ii) a stability check for elements in compression where the characteristic resistance relates to the buckling limit of the element. An element (member or plate) is checked at typical sections (at least both ends and mid-span) against resistance and buckling. This verification also includes the effect of water pressure on deep-water structures. Tubular joints are checked against punching under various load patterns. These checks may indicate the need for local reinforcement of the chord using over-thickness or internal ring-stiffeners. Elements should also be verified against fatigue, corrosion, temperature, or durability wherever relevant.

4.21.2 Allowable Stress Method

This method is presently specified by American codes. The loads remain unfactored and a unique coefficient is applied to the characteristic resistance to obtain allowable stress as shown in Table 4.9.

Normal and extreme conditions, respectively, represent the most severe conditions under which (a) the plant is to operate without shutdown and (b) the platform is to endure over its lifetime.

4.21.3 Limit State Method

This method of design is enforced by European and Norwegian authorities and has now been adopted by the American Petroleum Institute (API) as it offers a more

---

TABLE 4.9
Coefficient for Resistance to Stresses

Condition	Axial	Strong Axis Bending	Weak Axis Bending
Normal	0.60	0.66	0.75
Extreme	0.80	0.88	1.00

TABLE 4.10
Load Factors

Limit State	Load Categories				
	P	L	D	E	A
ULS (normal)	1.3	1.3	1.0	0.7	0.0
ULS (extreme)	1.0	1.0	1.0	1.3	0.0
FLS	0.0	0.0	0.0	1.0	0.0
PLS (accidental)	1.0	1.0	1.0	1.0	1.0
PLS (post-damage)	1.0	1.0	1.0	1.0	0.0
SLS	1.0	1.0	1.0	1.0	0.0

uniform reliability (Fjeld, 1977; Freudenthal and Gaither, 1969; Furnes, 1977). Partial factors are applied to the loads and the characteristic resistance of the element as given in Table 4.10. They reflect the amount of confidence placed in the design value of each parameter and the degree of risk accepted under a limit state as discussed below:

- Ultimate limit state (ULS), which corresponds to an ultimate event considering the structural resistance with appropriate reserve.
- Fatigue limit state (FLS), which relates to the possibility of failure under cyclic loading.

where P represents permanent loads (structural weight, dry equipment, ballast, hydrostatic pressure), L represents live loads (storage, personnel, liquid), D represents deformations (out-of-level supports, subsidence), E represents environmental loads (wave, current, wind, earthquake), A represents accidental load (dropped object, ship impact, blast, fire). The material partial factors for steel usually are taken equal to 1.15 for ULS and 1.00 for PLS and SLS design. Guidance for classifying typical conditions into typical limit states is given in Table 4.11.

The progressive collapse limit state (PLS) reflects the ability of the structure to resist collapse under accidental or abnormal conditions, while the serviceability limit state (SLS) corresponds to the criteria for regular use or durability (often specified by the plant operator).

The analysis of the offshore platform is an iterative process, which requires progressive adjustment of the member sizes concerning the forces they transmit until a safe and economical design is achieved. It is therefore of utmost importance to start the primary analysis from a model which is close to the final, optimized one. The simple rules given below provide an easy way of selecting realistic sizes for the main elements of offshore structures in moderate water depth (up to 80 m) where dynamic effects are negligible.

TABLE 4.11
Conditions Specified for Various Limit States

Conditions	P/L	E	D	A	Design Criterion
		Loadings			
Construction	P				ULS, TSLS
Load-out	P	Reduced wind	Support displacement		ULS
Transport	P	Transport wind and wave			ULS
Tow-out (accidental)	P			Flooded compartment	PLS
Launch	P				ULS
Lifting	P				ULS
In-Place (normal)	P + L	Wind, waves, and snow	Actual		ULS, SLS
In-Place (extreme)	P + L	Wind and 100-year wave	Actual		ULS, SLS
In-Place (exceptional)	P + L	Wind and 10,000-year wave	Actual		PLS
Earthquake	P + L	10^{-2} quake			ULS
Rare earthquake	P + L	10^{-4} quake			PLS
Explosion	P + L			Blast	PLS
Fire	P + L			Fire	PLS
Dropped object	P + L			Drill collar	PLS
Boat collision	P + L			Boat impact	PLS
Damaged structure	P + reduced L	Reduced wave and wind			PLS

Jacket Pile Sizes

- Calculate the vertical resultant (dead weight, live loads, and buoyancy), the overall shear, and the overturning moment (environmental forces) at the mudline.
- Assuming that the jacket behaves as a rigid body, derive the maximum axial and shear force at the top of the pile.
- Select a pile diameter by the expected leg diameter and the capacity of pile-driving equipment.
- Derive the penetration from the shaft friction and tip bearing diagrams.
- Assuming an equivalent soil subgrade modulus and full fixity at the base of the jacket, calculate the maximum moment in the pile and derive its wall thickness.

Deck Leg Sizes

- Adapt the diameter of the leg to that of the pile.
- Determine the effective length from the degree of fixity of the leg into the deck (depending upon the height of the cellar deck).
- Calculate the moment caused by wind loads on topsides and derive the appropriate thickness.

Jacket Bracings

- Select the diameter to obtain a span/diameter ratio between 30 and 40.
- Calculate the axial force in the brace from the overall shear and the local bending caused by the wave assuming partial or total end restraint.
- Derive the thickness such that the diameter/thickness ratio lies between 20 and 70 and eliminate any hydrostatic buckle tendency.

Deck Framing

- Select spacing between stiffeners (typically 500–800 mm).
- Derive the plate thickness from formulae accounting for local plastification under the wheel footprint of the design forklift truck.
- Determine by straight beam formulae the sizes of the main girders under "blanket" live loads and/or the respective weight of the heaviest equipment.

The static in-place analysis is the basic and generally the simplest of all analyses. The structure is modeled as it stands during its operational life and subjected to pseudo-static loads. This analysis is always carried out at the very early stage of the project, often from a simplified model, to size the main elements of the structure. The primary model should account for eccentricities and local reinforcements at the joints. For example, a typical model for the North Sea jacket may feature over 800 nodes and 40,00 members. The contribution of appurtenances like risers, J-tubes, caissons, conductors, boat fenders, etc., to the overall stiffness of the structure is normally neglected. They are, therefore, analyzed separately, and their reactions are applied as loads at the interfaces with the main structure. Since their behavior is nonlinear, foundations are often analyzed separately from the structural model. They are represented by an equivalent load-dependent secant stiffness matrix; coefficients are determined by an iterative process where the forces and displacements at the typical boundaries of structural and foundation models are equated. This matrix may need to be adjusted to the mean reaction corresponding to each loading condition. The static in-place analysis is performed under different conditions where the loads are approximated by their pseudo-static equivalent. The primary loads relevant to a given condition are multiplied by the appropriate load factors and combined to produce the most severe effect in each individual element of the structure. A dynamic analysis is normally mandatory for every offshore structure but can be restricted to the main modes in the case of stiff structures.

4.21.4   Fabrication and Installation Loads

These loads are temporary and arise during the fabrication and installation of the platform or its components. During fabrication, various structural components generate lifting forces, while in the installation phase forces are generated during platform load-out, transportation to the site, launching and upending, as well as during lifts related to installation. According to the Det Norske Veritas (DNV, 1982) rules, the return period for computing design environmental conditions for installation and fabrication loads is three times that of the duration of the corresponding phase. API-RP2A, on the other hand, leaves this design return period up to the owner, while the BS6235 rules recommend a minimum recurrence interval of 10 years for the design environmental loads associated with the transportation of the structure to the offshore site.

4.21.5   Lifting Force

Lifting forces are functions of different parameters; namely, i) weight of the structural component being lifted, ii) number and location of lifting eyes used for the lift, iii) angle between each sling and the vertical axis, and iv) conditions under which the lift is performed; details are illustrated in Figure 4.26. All members and connections of a lifted component must be designed for the forces resulting from the static equilibrium of the lifted weight and the sling tensions. Moreover, API-RP2A recommends that to compensate for any side movements, lifting eyes and the connections to the supporting structural members should be designed for the combined action of the static sling load and a horizontal force equal to 5% of this load, applied perpendicular to the member at the center of the pinhole. All these design forces are applied as static loads if the lifts are performed in the fabrication yard. If, however, the lifting derrick or the structure to be lifted is on a floating vessel, then dynamic load factors should be applied to the static lifting forces. A factor of 2 is applied for members and connections and 1.35 for all other secondary members. For load-out at sheltered locations, the corresponding minimum load factors for the two groups of structural components are 1.5 and 1.15, respectively.

4.21.6   Load-Out Force

These are forces generated when the jacket is loaded from the fabrication yard onto the barge. If the load-out is carried out by direct lift, then, unless the lifting arrangement is different from that to be used for installation, lifting forces need not be computed. This is because lifting in the open sea creates a more severe loading condition, which requires higher dynamic load factors. If load-out is done by skidding the structure onto the barge, several static loading conditions must be considered, with the jacket supported on its side. Such loading conditions arise from the different positions of the jacket during the load-out phases, as shown in Figure 2.26. Since the movement of the jacket is slow, all loading conditions can be taken as static. Typical values of friction coefficients for the calculation of skidding forces are (i) steel on

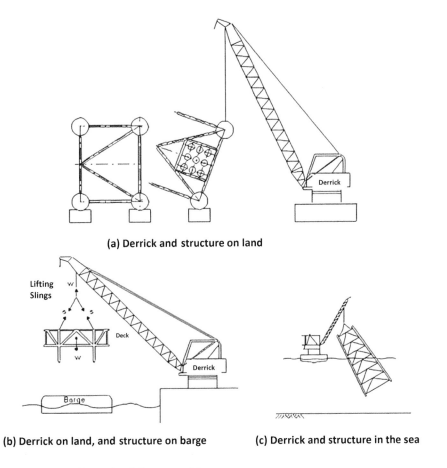

(a) Derrick and structure on land

(b) Derrick on land, and structure on barge

(c) Derrick and structure in the sea

FIGURE 4.26 Lifts under different conditions.

steel without lubrication (0.25); (ii) steel on steel with lubrication (0.15); (iii) steel on Teflon (0.10); and (iv) Teflon on Teflon (0.08).

4.21.7 TRANSPORTATION FORCES

These forces are generated when platform components (jacket, deck) are transported offshore on barges or self-floating. They depend upon the weight, geometry, and support conditions of the structure (by barge or by buoyancy) and also on the environmental conditions (waves, winds, and currents) that are encountered during transportation. The types of motion that a floating structure may experience are shown schematically in Figure 4.27. To minimize the associated risks and secure safe transport from the fabrication yard to the platform site, it is important to plan the operation carefully by considering the following (API-RP2A):

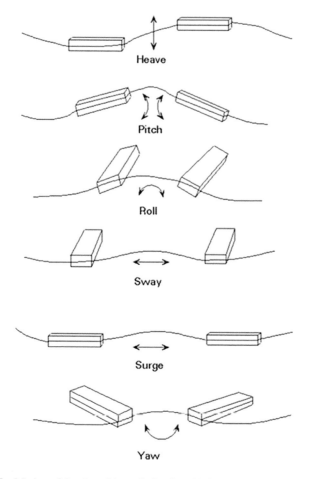

FIGURE 4.27 Motion of floating objects during installation.

- Previous experience along the tow route.
- Exposure time and reliability of predicted "weather windows."
- Accessibility of safe havens.
- Seasonal weather system.
- Appropriate return period for determining design wind, wave, and current conditions, taking into account the characteristics of the tow such as size, structure, sensitivity, and cost.

The motion of the tow—i.e., the structure and supporting barge—generates transportation forces. They are determined by the design of winds, waves, and currents. If the structure is self-floating, the loads are calculated directly. Vortex suppression systems help in controlling these forces (Chandrasekaran and Merin, 2016). According to API-RP2A, towing analyses must be based on the results of model basin tests or appropriate analytical methods and must consider wind and wave directions parallel, perpendicular, and at 45° to the tow axis. Inertial loads shall be computed from a rigid body analysis of the tow by combining roll and pitch with heave motions, when the size of the tow, magnitude of the sea state, and experience make such assumptions reasonable. For open sea conditions, typical values are 20° (for single amplitude roll motion) and 10° for single amplitude pitch motion. The period of roll or pitch is taken as 10 s, while heave acceleration is taken as 0.2 g. When transporting an oversized jacket by barge, stability against capsizing is a primary design consideration because of the high center of gravity of the jacket. Moreover, the relative stiffness of the jacket and barge may need to be taken into account, together with the wave slamming forces that could result during a heavy roll motion of the tow, as shown in Figure 4.28. Structural analyses are carried out for designing the tie-down braces and the jacket members affected by the induced loads.

4.21.8   LAUNCHING AND UPENDING

These forces are generated during the launch of a jacket from the barge into the sea and during the subsequent upending into its proper vertical position to rest on the seabed (BS6235, 1982). A schematic view of the five stages of the operation can be seen in Figure 4.29. Five stages in a launch-upending operation are (i) jacket slides along the skid beams; (ii) jacket rotates on the rocker's arms; (iii) jacket rotates and

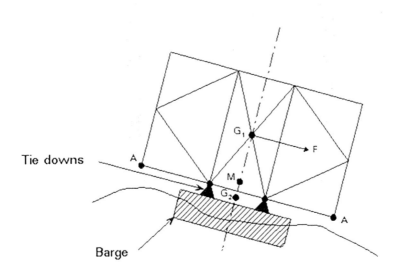

F   = Component of gravity plus inertia
G_1 = Centre of gravity of jacket
G_2 = Centre of gravity of the tow
M   = Metacentre of the tow
A   = Areas of potential impact

FIGURE 4.28   View of launch barge and jacket undergoing motion.

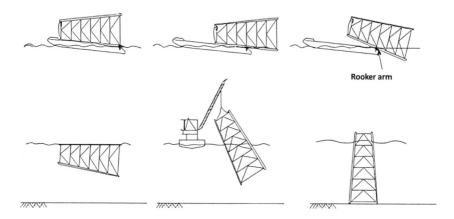

FIGURE 4.29 Launching and upending.

slides simultaneously; (iv) detaches entirely and comes to its floating equilibrium position; and (v) jacket is upended by a combination of controlled flooding and simultaneous lifting by a derrick barge. Both the static and dynamic loads for each stage of the above under the action of wind, waves, and current need to be included in the analysis. To start the launch, the barge must be ballasted to an appropriate draft and trim angle, and subsequently, the jacket must be pulled towards the stern by a winch. Sliding of the jacket starts as soon as the downward force (gravity component and winch pull) exceeds the friction force. As the jacket slides, its weight is supported on the two legs that are part of the launch trusses. The support length keeps decreasing and reaches a minimum equal to the length of the rocker beams when rotation starts. It is generally at this instant that the most severe launching forces develop as reactions to the weight of the jacket. During the last two stages, variable hydrostatic forces arise, which have to be considered by all members affected. Buoyancy calculations are required for every stage of the operation to ensure fully controlled, stable motion. Computer programs are available to perform the stress analyses required for launching and upending and also to portray the whole operation graphically.

4.21.9 ACCIDENTAL LOAD

According to the DNV rules, accidental loads are ill-defined concerning intensity and frequency, which may occur as a result of an accident or exceptional circumstances. Examples of accidental loads are loads due to collision with vessels, fire or explosion, dropped objects, and unintended flooding of buoyancy tanks. Special measures are typically taken to reduce the risk of accidental loads. For example, protection of wellheads or other critical equipment from a dropped object can be provided by specially designed, impact-resistant covers. An accidental load can be disregarded if its annual probability of occurrence is less than 10^{-4}. This number is the estimate of order of magnitude and is extremely difficult to compute.

4.22 CORROSION

Corrosion is the deterioration of material by chemical interaction with the environment. This term also refers to the degradation of plastics, concrete, and wood, but generally refers to metals. The corrosion process produces a new and less desirable material from the original metal, which results in a loss of function of the component or system. The common product of corrosion is "rust," which is formed on the steel surface. The basic corrosion cell is shown in Figure 4.30. The basic corrosion cell needs three components: an anode, a cathode, and an electrolyte medium.

Electron flow during a corrosion process is shown in Figure 4.31. There will be a measurable, direct current (DC) voltage, which can be read in the metallic path between the anode and the cathode. When both the anode and the cathode are electrically bonded, the anode is positively charged, and the cathode is negatively charged. Conventional current flows from positive to negative, and thus, current discharges from the anode and is picked up at the cathode through the electrolyte. Current returns from the cathode to the anode through an electrical path. This flow has a detrimental effect on the anode known as "corrosion." It is important to note that corrosion occurs at the anode and not at the cathode. Corrosion is a process in which ions are involved. For corrosion to take place, three basic requirements are necessary: (1) a medium to move, which is water in the case of ocean structures as the members are continuously exposed to seawater; (2) oxygen to activate the process, which is present in abundance; and (3) a metal, which should be willing to give up electrons to start the process. The corrosion process results in the formation of a new material, which may react again or could be protective of the original metal. The anode and cathode in a corrosion process may be on two different metals connected forming a

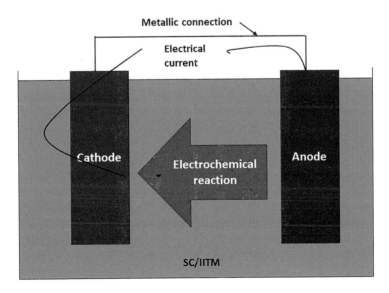

FIGURE 4.30 Basic corrosion cell.

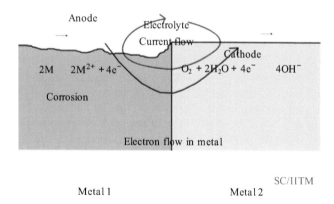

FIGURE 4.31   Electron flow during corrosion process.

bimetallic couple (galvanic couple), or, as in the case of rusting of steel, they may be formed on the metal surface.

4.22.1   CORROSION IN STEEL

Steel is the basic material of construction in the offshore industry. Corrosion, to a large extent, is governed by the oxygen content of seawater. The corrosion rate of steel in the marine environment is related to the rate at which a ferrous corrosion product is leached or washed from the film of rust. When one of the products of corrosion becomes soluble, the formation of a protective barrier film becomes impossible. The presence of copper and nickel, even in small quantities in the low alloy steel, enhances their corrosion resistance by altering the structure of the barrier film formation. They help to produce a tighter, denser barrier film with less of a tendency to be removed by leaching or spalling. Figure 4.32 shows a typical offshore platform with different corrosion zones marked. The figure shows different regions at which the corrosion takes place in an offshore platform. The top zone is the atmospheric zone where derricks and deck modules are located. This zone experiences the minimum rate of corrosion because the members are not in direct contact with water. Hence, leaching or washing of the thin barrier film is at a lower probability.

Atmospheric sea exposure is always present on the top portion of the topside where derrick and deck modules are present. It contains generally precipitated salt, and the condensation process takes place in this region; a soffit of the deck slab is the most vulnerable candidate for corrosion in this region. The corrosion rate of steel in the marine atmosphere is related to the rate at which the ferrous corrosion product is leached or washed off from film or rust. A protective barrier film is being created on the top of the member, as a by-product of the corrosion process. If there is a possibility that this barrier film can be leached off or washed off, then the corrosion process can be activated. In the case of an atmospheric zone, there is a low possibility that this film can be washed off. When one of the products of corrosion becomes soluble, the formation of a protective barrier film is impossible.

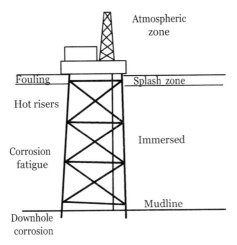

FIGURE 4.32   Offshore jacket platform with various corrosion zones.

Severe corrosion is seen in the splash zone due to continuous wetting and drying because of tidal variations in the sea. It results in pitting corrosion in the tidal area. Due to continuous contact of seawater with a higher lateral force, a thin barrier film, even if formed, will be washed away immediately; this expedites corrosion in the splash zone. Rust films in this zone have little opportunity to become dry as this zone is subjected to alternate wetting and drying continuously. Even though the rust films may be formed, they will be leached off automatically. This is aggravated because of the presence of abundant oxygen content in this region. The rate of corrosion in this splash zone is several times greater than that of the continuous immersion part of the member. It is interesting to note that the same member (say, e.g., jacket legs) passes through different regions of corrosion, resulting in the development of a bimetallic couple. One part of the member becomes anodic and the other cathodic; the presence of an electrolyte activates the corrosion process very fast. Therefore, the rate of corrosion in this region is seen as several times higher than that of the other regions, which are continuously immersed in seawater.

In the tidal zone, corrosion reaches a minimum because of the protective action of oxygen concentration cell currents present in this region. Steel surface in a tidal zone is in contact with highly aerated seawater and therefore becomes cathodic. As only the anodic part of the member corrodes, the adjacent submerged surface where the oxygen content is less becomes anodic, and therefore, it gets corroded severely. For example, the members that are covered with oxygen-shielding organisms like marine growth may get less oxygen content on the surface, and they become anodic and corrode faster than the members present in the tidal zone. The current flows from the anode, which is the submerged surface, to the cathode, which is a tidal zone in this area of the sea environment. This enables sufficient cathodic protection to the members in the tidal zone automatically. This is caused by the differential aeration or formation of marine growth in the regions that are immersed below.

In the immersed zone where jackets and mud lines are present, corrosion is reduced due to the decrease in oxygen concentration. In this zone, corrosion is principally governed by the rate of diffusion of oxygen through layers of rust and marine organisms. The corrosion rate is not influenced by the seawater temperature and tidal velocity. Kindly note that the rate of corrosion is not determined by the temperature gradient in seawater with the increase in water depth. The corrosion rate may go up in the vicinity of the mud line, but further down, it is much less. This is due to the presence of marine organisms, which can generate additional concentration cells and sulfur compounds in the vicinity of the mud line. Due to this, they become anodic compared to the remaining part and get corroded. The corrosion rate reduces well below the mud line because of the lower availability of dissolved oxygen content. Barrier films, once they are formed in this region, are relatively undisturbed. Therefore, they form a protective coating automatically, and that protects the members in this region.

Various factors influence the corrosion rate of steel members in offshore structures. Considering the effect of current velocity on the corrosion rate, it is understood that any gentle motion will not affect or disturb the formation of the protective barrier on members. However, when the platform motion is larger, the barrier layer formation becomes thinner and is easily broken under higher current velocity. Therefore, current velocity plays an important role in accelerating the corrosion rate; an increase in current velocity increases the rate of corrosion. Considering different methods of corrosion protection, the majority of the members of the platform present in the atmospheric and tidal zones can be protected by painting. For those present in the splash zone, special methods such as providing extra steel, Monel wrapping, or sheathing are recommended as a corrosion protection measure. In the case of the immersion zone, sacrificial anode technique or cathodic protection methods are employed. In the mud line below, no special methods of corrosion protection are advocated, as there is a small amount of corrosion.

The effect of water depth on the corrosion rate is also an interesting viewpoint. At a water depth of greater than 1800 m, the temperature drops to less than 4°C, in comparison with 24°C at the surface. This results in a substantial reduction in the corrosion rate. Metal surfaces are relatively free of marine biofouling below 700 m. Dissolved oxygen drops along the depth but rises again below 820 m. This rise is significant, in comparison with the surface concentration. Ocean layers are not homogeneous; various layers are differentiated by different oxygen and salinity contents. Corrosion decreases at greater water depth as the temperature decreases. Fouling and pitting associated with fouling also tend to decrease. Mooring lines, which extend to different zones of corrosion, face a critical problem. Due to part of the mooring line becoming cathodic, corrosion is set at the local level along the length of the mooring. This is called the "long-line effect." Mainly due to the typical oxygen cell concentration attack, long length of mooring lines is subjected to different layers of varied oxygen concentration. This alters the rate of corrosion in different segments along the length of the mooring line, which is a long-line effect. Galvanized mooring lines are common candidates for such problems.

4.22.2   Corrosion in Concrete

Concrete, which has embedded steel, has a high degree of protection against corrosion. As concrete is alkaline in nature, it provides barrier protection to steel reinforcement. The presence of chloride in sufficient quantities in the vicinity of steel results in cracking, spalling, and delamination of concrete. Early detection of the corrosion activity can assist in planning corrosion preventive measures. Chloride-induced corrosion is the most serious cause of deterioration in RCC structures. Structural weakening caused by corrosion can reduce its service life by 20 years. The corrosion rate is accelerated in RCC members whenever there is exposure to the source of chloride. Patching of damaged areas does not stop corroding but spreads to other areas faster. The corrosion mechanism in RCC structures can be easily understood: Steel is in a passive state in concrete. If chlorides reach the steel surface by ingression, this passive layer is broken. This initiates the corrosion process. Corrosion current flows from one part of the reinforcement (anode) to another part (cathode). Because of this current flow, steel corrodes at the anode and produces rust. As a result, reinforcing steel develops a tendency to revert to its natural oxide state, which is not capable of withstanding the encountered stresses. Corroded steel can expand four to five times that of its normal volume. This will result in cracking, spalling, and delamination of concrete. This further exposes more steel contact areas for chloride ingression and accelerates corrosion.

4.22.3   Realkalization

Fresh concrete has inherent alkalinity, which provides passive protection to steel. Ingression of carbon dioxide creates carbonated concrete with lower alkalinity, which results in loss of passive protection to rebar. It also accelerates the corrosion of steel reinforcement. Realkalization involves an electrochemical technique of passing sustained low-voltage current between the temporary anodes on the surface of concrete and steel reinforcement. The period of application can vary from three to seven days. Electrolyte covering is done by spraying cellulose fiber saturated in sodium carbonate solution. Surface nodes, embedded in alkali-rich paste, draw alkali into concrete through rebar. Realkalization takes place in concrete to initiate the formation of a natural protective oxide film over rebar.

4.23   CORROSION PREVENTION

There are different ways by which corrosion can be prevented: i) conditioning the metal surface, ii) conditioning the corrosive environment, iii) controlling the electrochemical reaction, which is responsible for corrosion, iv) fighting corrosion with corrosion, v) coating the metal, and vi) alloying the metal. The rate of corrosion can be reduced by retarding either the anodic or the cathodic reaction. The principle aim behind any corrosion prevention or protection method is to fight corrosion with corrosion. It means that, to reduce corrosion, create an additional member and allow it to corrode. Sacrificing the additional members can prevent corrosion on the existing members of the structure. This is called the sacrificial anode method. Using another metal to coat an existing metal surface is the common case in zinc or tin coating. It is generally applied on steel as an external coating surface. A protective coating derived from the metal surface itself can also be applied. For example, the metal surface of a member can be coated with aluminum oxide; organic coatings such as resins, plastics, paints, enamel, oil, and greases are also used. Coating a metal can also help reduce the corrosion rate but pose a serious threat to the sea environment. Alternatively, one can also alloy the metal to produce a corrosion-resistant alloy. A classic example is stainless steel, which is ordinary steel alloyed with nickel and chromium. By conditioning the corrosive environment, one can control corrosion. Oxygen is one of the main components required to activate the corrosion process. Removal of oxygen can help to retard the rate of corrosion. Removal can be achieved by adding strong-reducing agents. For example, sulfites can reduce the presence of oxygen content in the sea environment. Removal of oxygen is not advisable in the open environment because of the presence of oxygen in abundance.

4.23.1 Corrosion Protection

There are many methods to protect offshore structures from corrosion. The atmospheric zone is one where corrosion is not very severe but is marginally high. One can use coatings to protect the members. In the splash zone, where the corrosion rate is very severe, one can use Monel-400 or other metal cladding. Monel-400, an alloy of 18 gauge thick (approximately equal to 1.02 mm) is attached to the tubular member in the splash zone. This is done either by bonding the Monel sheathing on the parent member or by welding. Monel-400 has a high modulus of elasticity and will not get damaged under the stress conditions that are caused by this installation process. However, it is likely to get damaged by tearing or impact forces as the sheathing is too thin. There is a tendency for the sheathing to peel off or tear off from the parent surface of the material. Alternatively, austenitic stainless steel 304, which is an alloy of chromium-nickel stainless steel, can be wrapped over the surface of the members in the splash zone. The advantages of these applications are a high degree of weldability and increased stiffness.

Another alternative material, which is also commonly deployed in the splash zone, is a copper-nickel alloy of either 70%–30% or even 90%–10% composition. This adds stiffness to the members during installation. One can also use steel wear plates of 6–13 mm thickness. They also add strength and stiffness to the members and improve their resistance against impact loads. This application is more common in the arctic regions, where the temperature variations can be very large. Splashtron and vulcanized neoprene are the two varieties of rubber products that can be used as a sheathing layer on the members near the splash zone. Splashtron is an elastomeric rubber sheathing that is braced to the members. It is highly resistant to corrosion and mechanical abuse. It has very high tearing strength when it is hardened. It adheres to the parent material very firmly and becomes more or less homogeneous in action with that of the parent material. Thickness usually varies from 5 to 13 mm, which is high in comparison with that of the Monel sheathing.

The use of corrosion inhibitors is also one of the effective methods of reducing corrosion. Corrosion inhibitors are of different types: anodic, cathodic, adsorption, and mixed. Corrosion inhibitors are other alternatives for corrosion protection of members of ocean structures. These are chemical additives that, when added to the corrosive aqueous environment, interfere with the chemical reaction and reduce the rate of corrosion. Anodic inhibitors inhibit the response that takes place at anodes. They suppress the cathodic reactions that occur in a bimetallic couple. Adsorption-type corrosion inhibitors generally form a film on the surface of the member; they physically block the surface from the corrosive environment. Corrosion inhibitors are commonly deployed in deepwater platforms in the immersion and splash zones. Anodic inhibitors are more popular among all of these three types of corrosion inhibitors.

Another effective method is to control the electrochemical reaction responsible for the corrosion process. This technique is done by passing an anodic or cathodic current inside the metal. Cathodic protection is an important and common method of corrosion protection in the marine environment. In principle, it can be applied to any metallic surface that is in contact with the bulk electrolyte. This condition is automatically fulfilled in the case of offshore structures as seawater with impurities of sulfites and chemicals acts as the electrolyte. This is advantageous for members buried in soil or immersed in water and hence cannot be applied in the splash and atmospheric zones; alternate wetting and drying conditions are not suitable for this kind of corrosion protection measure. Figure 4.33 shows a schematic view of anodic and cathodic reactions on a metal surface.

The figure shows the formation of a bimetallic couple in the presence of a bulk electrolyte. The anodic reaction releases an electron and becomes positive, whereas a cathodic reaction receives an electron and becomes negative. Therefore, the anodic part is continuously corroded, but the cathodic part is protected; hence, the name cathodic protection. To protect the parent member, one should provide another material as an anode, which is capable of forming a bimetallic couple with the parent metal; in this case, the anode is sacrificed. Cathodic protection can be achieved in two ways: (1) by using galvanic anodes, termed the sacrificial anode technique, and (2) by the impressed current method. Metal to be protected is connected as a cathode, whereas an external metal is connected as an anode. When connected to an electric DC, the anode is ready to release electrons in a chemical reaction. The external metal

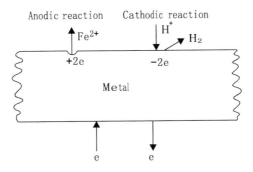

FIGURE 4.33 Cathodic and anodic reactions on metal surface.

that is provided may be a galvanic anode, where the current results from the potential difference between two metals. It forms a galvanic couple as well; alternatively, the current is impressed on the metal by an external DC power source.

The galvanic anode systems employ reactive metals as auxiliary anodes that are directly connected to the metal, which is to be protected. Therefore, the member or the steel surface, which is to be protected, should be made as a cathode. An additional member is introduced to act as a galvanic anode deliberately. The potential difference between the anode and the parent steel, as indicated by their respective positions in the electrochemical series, will make the new material anodic; corrosion is initiated in the presence of an electrolyte. The current flows from the anode to the parent metal, which results in the corrosion of the anode. Thus, the whole surface of steel, which is now cathodic, is protected as it is negatively charged. This is termed as "cathodic protection." As a new metal provided acts as an anode that corrodes, this is also termed a sacrificial anode technique of cathodic protection. Metals that are commonly used as sacrificial anodes are aluminum, zinc, and magnesium. They are used in the form of rods, big blocks, or wires that can be wounded around the members. Big blocks can either be bolted or welded to the structure. This system is advantageous as it is very simple to install and requires no external source power.

Localized protection is highly effective and immediately available on float-out. Moreover, this has less interaction with neighboring structures. One of the main disadvantages is that the current output available is relatively small. Therefore, monitoring a galvanic system for adequate corrosion protection is very difficult under surveys. A monitor system is highly sensitive to recording minor variations in voltage. The flow of electrons depends on the electrical resistivity of the electrolyte. A change in the structure—say, for example, deterioration of coatings—demands more current and hence more sacrificial anodes; therefore, sometimes, this method proves to be expensive. An alternate method by which one can also use cathodic protection is bypassing the impressed current on the metal. Impressed current systems employ either zero or low dissolution anodes. They use an external DC power to impress the current from an external anode onto the cathodic surface. Connections are similar to cathodic protection and are commonly applied to metallic storage tanks and RCC ocean structures. This is used as corrosion protection for members in the immersion zone.

Figure 4.34 shows a schematic view of the impressed current method as applied for a steel pipeline. Anodes are externally connected to the remote pipe. The current flows from the anode to the pipeline through earth or water, the bulk electrolyte (medium). Impressed current is passed at the location where the pipeline is laid; as it remains cathodic, it is protected. This method is effective only when the members are fully immersed in the medium; this method requires the contact of bulk electrolytes to activate the process. One of the main advantages of cathodic protection over other forms of anticorrosion treatments is its effectiveness in continuous monitoring. This is possible by maintaining a DC circuit. One can record the amount of electronic flow between the anodic and cathodic terminals, which is the index to measure the effectiveness of the treatment. Cathodic protection is commonly applied to surface-coated members, and there is a high probability of this coating being damaged. For example, members in the atmospheric or splash zone have a tendency for coatings to get washed or leached off due to the chemicals present in the sea environment.

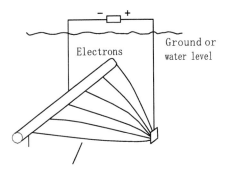

FIGURE 4.34 Impressed current method of cathodic protection.

The impressive current method has few merits. As it can supply a relatively more significant current compared to galvanic systems, effective monitoring of the control mechanism is highly feasible by impressing the current mechanism. It can provide high DC driving voltages and can be used in most types of electrolytes. As it can provide a flexible output, it may accommodate respective changes in the structural members. However, the system has some demerits. For example, intensive care should be taken to minimize the interaction with other structures. As it is uniformly available for larger protection surfaces, interaction between the structural members in elements is also highly feasible. Regular maintenance or monitoring is very important in this kind of protection system.

Cathodic protection is a common phenomenon implemented in the design stage of ocean structures. Exterior surfaces of ocean structures are protected by cathodic protection. Examples are pipelines, hulls of ships, bases of storage tanks, jetties and harbor structures, tubular joints in jacket structures, and foundation piles (Dziubinski et al., 2006). Floating offshore platforms and subsea structures are typical examples where cathodic protection has recently been very largely deployed. In the North Sea, the galvanic protection method against large uncoated platforms is found to be very cost-effective. Because coating maintenance costs are very high, offshore engineers prefer to use galvanic protection techniques in the Gulf of Mexico or the North Sea for most platforms. Galvanic systems are easy to install and are robust systems. As it requires no external power source, it is considered one of the main advantages.

Moreover, it provides protection immediately on float-out of the structure; this method of corrosion protection is instantaneous. Cathodic protection is also used to protect the internal surface of large-diameter pipelines and ballast tanks in ships. The inner surface of large oil storage tanks is also a common candidate for this method of corrosion protection. In a process industry that uses continuous circulation of coolant or water under differential temperatures, cathodic protection is the most preferred method to control the corrosion rate; if not, it can be completely prevented.

The cathodic protection system has specific requirements. This can be applied to members that are in contact with bulk electrolytes. In addition, a galvanic system requires a sacrificial anode, which should be directly welded to the structure; alternatively, a conductor can connect the anode to the structure. A secured connection with a minimum resistance between the conductor and the structure is also to be ensured. An impressed current system requires inert anodes, which are clusters of anodes connected, often in a backfill. It also requires an external DC power source and an electrically well-insulated system to ensure minimum resistance and a secure connection between the anodes, conductors, and the power source. The source of DC power, which is vital in the case of the impressed current method, can be ensured by using rectifiers of transformer units in conjunction with an existing AC supply; alternatively, one can use either diesel- or gas-driven alternators. In remote areas, power sources include thermoelectric generators and solar or wind generators for generating the required DC power for impressed current.

4.24   REPAIR AND REHABILITATION

Materials for the repair and rehabilitation of ocean structures are not under the recommendation of international codes. These codes only suggest repair procedures and desirable characteristics of materials for repair. Among several reasons for this limitation, the foremost is that the material choice for repair is case-specific (Zhang and Aktan, 1995). Ocean structures are constructed for a variety of functional requirements, which are very specific to the type of the chosen structural system. Repair of ocean structures is required to be carried out without affecting their functional routine. Furthermore, they cannot be relieved from the encountered environmental loads during repair. This means that ocean structures need to undergo repair, although they are under the influence of various environmental loads, which is an important challenge. The characteristics of a material chosen for repair should enable speedy construction and attain the desired strength at the earliest possible time. This is because the downtime available for the repair of ocean structures is generally for a limited period; constraints may arise from the weather window or functional priorities. Moreover, repairing ocean structures is not preventive in general but only prescriptive to functional failure. In such cases, special issues related to their survivability under critical load combinations encountered by them are very critical issues. These types of structures cannot be dismantled or reconstructed but only be repaired. There are instances where an extensive repair needs to be carried out underwater (see, for example, the details of the repair of a ship dockyard in Pennsylvania, as shown in Figure 4.35). Repair carried out on the dockyard in recent times involved a new approach of supporting the deck on a new set of piles, although the dockyard was in service. Repair of the Cape May Ferry berthing Jetty in Cape May, New Jersey, enhanced the ferry-handling capacity of the Jetty (Figure 4.35). Designed and constructed new boardwalks using prestress concrete bulkheads are also equipped with state-of-the-art fender systems. Details of the repair works carried out in Exelon Power Corporation, Philadelphia, showed that a new set of steel auxiliary piles was installed to replace the deficient piles on the front end (Figure 4.36).

From the foregoing examples, it is clear that the repair of ocean structures is a state-of-the-art procedure due to the updated demand for functional characteristics and enhanced load-carrying capacity. Reduction of forces are also feasible with perforated outer cover on main members. Studies showed effective reduction of forces (Wang and Ren, 1993, 1994; Mutlu Sumer and Freddie, 2003; Losada et al., 1995).

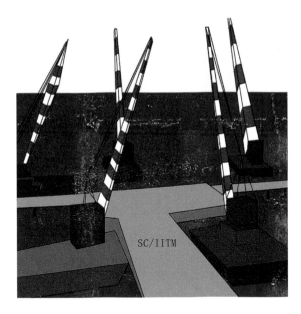

FIGURE 4.35   Ship dockyard, Pennsylvania.

FIGURE 4.36   Upgradation of Cape May, New Jersey.

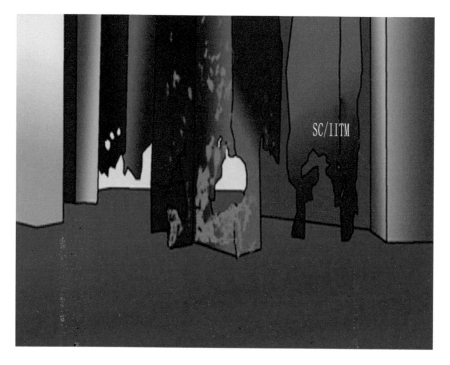

FIGURE 4.37   Steel auxiliary piles, replacing damaged piles.

Hence, the material chosen for repair is not based on the existing design requirements but should also compensate for the degraded performance of deteriorated materials. Both the factors—namely, strength and serviceability—must be fulfilled. Perforated covers around members will help reducing the forces without any additional strength from the materials (Kenny et al., 1976; Chandrasekaran and Parameswara, 2011; Chandrasekaran et al., 2014; Chandrasekaran and Madhavi, 2014a, 2014b, 2014c, 2014d, 2015a, 2015b, 2015c, 2015d, 2015e).

Before the actual repair can be carried out, the following factors need to be established: (1) the existing strength of the structure, (2) the magnitude of the proposed repair, (3) the cost factor, (4) the shutdown time of the service of the structure, and (5) feasibility of the proposed repair work. If all the foregoing factors are included in the study based on which repair methodology is suggested, then the study is termed an "integrity analysis." Repair of ocean structures is full of challenges. Unlike land-based structures, ocean structures need to be repaired in a hostile environment. It requires a set of specialized equipment, chemicals, and construction expertise to carry out such repairs. It also requires state-of-the-art electronic systems to map underwater conditions before and after repair. These equipment studies include hydrographic survey equipment, side-scan sonar imaging, instruments to measure the ultrasonic thickness of steel members, underwater photography/video, and marine borer assessment.

Repair of ocean structures also poses a set of unique challenges; the foremost is that the structure has to remain in service during repair. Therefore, the load-carrying capacity should not be challenged when repairs are being attempted on ocean structures. Specialized methods and equipment are generally used for two reasons: (1) to minimize the shutdown time of the structure during repair due to limited availability of time for repairing, and (2) to minimize the damage to existing structures during repair. Other factors are as follows: (1) the repair process should also be cost-effective, and (2) a long-term solution is demanded. It is not because such repairs need to be evaluated from an economic perspective, but for a valid reason that ocean structures cannot be intervened for repair frequently. As preventive maintenance is not a usual practice in many ocean structures commissioned worldwide, repair processes become more complicated as they are generally requested only in an emergency; enough time is not available for detailed studies and verification.

Hence, offshore engineers should thoroughly understand various repair methodologies and chemicals available to carry out repairs in emergencies. Unlike land-based structures, ocean structures need to be repaired in hostile conditions as structures mostly have less or remote access to land. They require specialized equipment, chemicals, and construction expertise to repair ocean structures. They also require state-of-the-art electronic systems to map the underwater conditions of the ocean structures. They include hydrographic survey equipment, site scanners, and sonar imaging equipment. Underwater videography, photography, and marine borer assessment are the common methods used during the repair process. Repair processes are not generally prescribed in the standard literature and are not generally recommended by international codes. This is because various chemical admixtures that are generally used for repairs are case-specific.

5 Petroleum Production

Summary

This chapter deals with the mechanics involved in petroleum production. It explains details of types of drilling rigs and their working principles to impose a better understanding of their mechanisms. A short description about blowout and kick is helpful to understand the reasons of their occurrence and methods to avoid or control their risk occurrence. Petroleum production process is also presented with an introduction to oil and gas exploration and drilling, production, and process platforms. A brief section of oil recovery helps understanding the basics of EOR and thermal recovery methods. A brief note on oil spill consequences and prevention helps readers to learn this important segment of environmental pollution arising from oil and gas exploration.

5.1 PETROLEUM DRILLING

One of the initial steps followed by the operating company for drilling operations is to prepare a drilling proposal followed by the drilling program. Reservoir engineers and geologists from the operating company are involved in the preparation of the drilling proposal. The information provided in the drilling proposal is used to design the drilling program. The drilling proposal contains information like target depth, target location, cross-section, well objective, and also the prediction of pore pressure profile (Austin, 2012). Drilling engineers preparing the drilling program take into account, the type of drilling rig, location, casing and hole size, type of drilling fluid, directional profile, drill bit, well control equipment, and also the hydraulics program (Krishna et al., 2018). In the early 18th century, for drilling activities, cable tool drilling was used. This technology was followed by rotary drilling rigs. In a rotary drilling rig, a drill bit is attached at the end of a hollow pipe. As the pipe rotates, drilling fluid is injected through the pipe and circulated to bring the rock cuttings to the surface. At offshore locations, mobile offshore drilling units (MODU) are used to carry out the drilling operations. Jack-up rigs and drilling barges are used for shallow-water drilling operations. For deep water and ultra-deepwater locations, semi-submersibles and drillships are used to carry out the drilling operations.

5.2 ROTARY DRILLING RIG

The rotary drilling rig, as shown in Figure 5.1 comprises several components and systems. The systems and components in the rig can be classified into six subsystems. The six subsystems are: a) rotary system, b) hoisting system, c) block and tackle system, d) circulating system, e) power system, f) well control system, and g) well monitoring system.

DOI: 10.1201/9781003497660-5

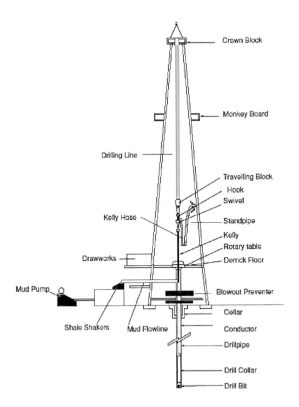

FIGURE 5.1 Rotary drilling rig.

5.2.1 ROTARY SYSTEM

A rotary system plays a crucial role in the drilling process as it enables the rotation of the drill string and the drill bit at the bottom of the borehole. It encompasses equipment that is necessary for achieving this rotation. At the top of the drill string, we have the swivel that serves three important functions. First, it supports the weight of the drill string. Second, it allows the string to rotate. And third, it permits the pumping of mud while the string is in motion. To ensure the proper functioning of the rotary system, the hook of the traveling block is securely attached to the bail of the swivel as shown in Figure 5.2. Additionally, the Kelly hose is connected to the gooseneck of the swivel. The Kelly, which is the initial section of pipe below the swivel, typically measures around 40 feet in length and features an outer hexagonal shape. This hexagonal (or sometimes square) shape is vital for transmitting rotation from the rotary table to the drill string. The Kelly has a right-hand thread connection at its lower end (pin) and a left-hand thread connection at its upper end (box).

To prevent excessive wear and tear on Kelly's connection threads during continuous drilling operations, a cost-effective component, known as a Kelly saver sub, is utilized between the Kelly and the first joint of the drill pipe. Moreover, to isolate high pressures and prevent backflow from the well in case of any influx at the bottom,

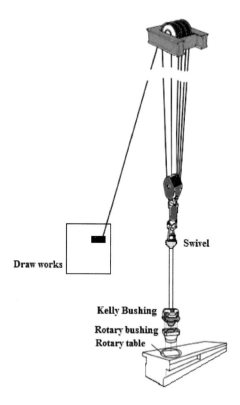

FIGURE 5.2   Rotary system.

valves called Kelly cocks are installed at both ends of the Kelly. The rotary table, positioned on the drill floor, allows for bi-directional rotation (both clockwise and anti-clockwise). The drillers can control the rotation from their console. This rotating table is designed with a square recess and four post holes. Furthermore, to safeguard the rotary table, a large cylindrical sleeve called a master bushing is employed.

The rotary table power requirements are calculated using the following relationship:

$$P_{rt} = \omega T/2\pi \qquad (5.1)$$

where P_{rt} is the power (hp), ω is the rotary speed (rpm), and T is the torque (ft-lbs). Slips are utilized to secure the suspended pipe firmly in place on the rotary table during the process of establishing or dismantling connections. Consisting of three hinged segments that taper and wrap around the upper section of the drill pipe, these slips allow the drill pipe to remain suspended from the rotary table while the top connection is being manipulated. The interior of the slips features a serrated surface designed to firmly grip the pipe and ensure stability. To disengage or "break" a connection, two large wrenches or tongs are employed. A set of pipes, consisting of three sections of drill pipe, is raised into the derrick until the lowermost part of the drill pipe becomes visible above the rotary table. At this juncture, the slips are inserted into the gap between the drill pipe and the master bushing in the rotary table to secure and support the remaining portion of the drill string. Subsequently, the breakout tongs are affixed to the pipe above the connection, while the make-up tongs are attached below the connection. By maintaining the make-up tong in position, the break-out tong is operated by the driller to disconnect the connection.

To establish a connection, the make-up tong is positioned above and the break-out tong, below the connection. In this scenario, the breakout tong remains stationary while the driller applies force to the make-up tong until the connection is securely fastened. Although tongs are typically employed to loosen or tighten connections to the desired torque, alternative methods exist for assembling connections before applying torque. For instance, when assembling the Kelly, a tong secures the lower tool joint while the Kelly is rotated using a Kelly spinner—a machine powered by compressed air. Additionally, a drill pipe spinner, or power tongs, may be utilized to assemble or disassemble connections using compressed air for power. A chain tong is preferred while assembling specific subs or specialized tools like MWD subs.

5.2.2   Hoisting System

The hoisting system, akin to a large pulley arrangement, is vital for lifting and lowering equipment in and out of the well; notably, for managing the drill string and casing. It encompasses the draw works, featuring a spinning drum wound with a wire rope, overseen by the driller using a clutch and brake system. The wire rope runs over sheaves at the top of the derrick (crown block) and down to another set of sheaves (traveling block), as shown in Figure 5.3. Suspended from the lower set of sheaves, a hook holds the drill string, while clamps known as elevators are affixed for maneuvering the drill string or casing. One end of the wire rope is secured below the rig floor (deadline), and the other end is wound onto the draw works (fast line). Multiple loops of the wire rope around the sheaves ensure stability, with the system's strength and configuration tailored to the load it needs to handle.

5.2.3   Block and Tackle System: A

"Block" refers to a group of pulleys or sheaves arranged on a single frame. When a rope is threaded through these pulleys, it forms an assembly called a tackle. The act of threading ropes or cables through these blocks is known as "reeving," and when completed, the block and tackle assembly is considered "rove." This block-and-tackle system effectively increases the tension force in the rope, enabling the lifting of heavy loads. They are commonly found on boats and sailing ships, where tasks are often done manually, as well as on cranes and drilling rigs, where heavy equipment takes over tasks once the block and tackle are set up. A derrick is a device equipped with a tackle system attached to the end of a beam for lifting and lowering objects. A derrick is constructed as a framework or tower made of wood or steel, positioned over deep oil well drill holes. It supports the tackle for drilling, facilitates the raising and lowering of drilling tools, and assists in inserting and removing well casing or

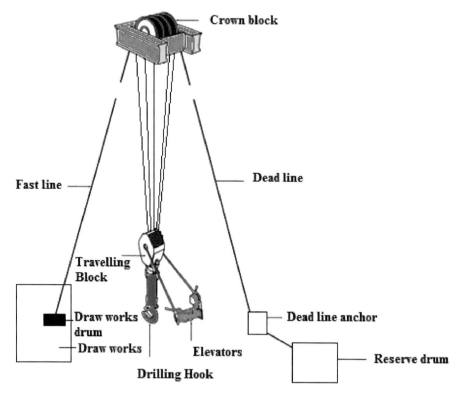

FIGURE 5.3 Hoisting system.

pipe. Smaller versions mounted on trucks are also referred to as derricks. Essentially, a derrick functions as a type of crane.

5.2.4 CIRCULATING SYSTEM

Drilling fluid is circulated through the drill string and up the annulus. The rock cuttings formed after the drilling process are brought to the surface with the support of the drilling fluids as shown in Figure 5.4.

Drilling fluids—otherwise referred to as drilling mud—are usually a mixture of water, bentonite, barite, which acts like a weighing material, and additives. Drilling fluids have several functions during the process of drilling. The major role of the drilling fluid is to retrieve the cuttings formed during the drilling process and also to exert a hydrostatic pressure on the formation below to prevent entry of the reservoir fluids to the wellbore. The drilling cuttings from the drilling mud are separated using solid removal equipment and are circulated back. Periodic checks are carried out on the drilling mud to ensure that the desired properties are maintained. To deliver the drilling mud in high pressure and volume, positive displacement triplex or duplex pumps are used. To drive the several subsystems, the rig requires power.

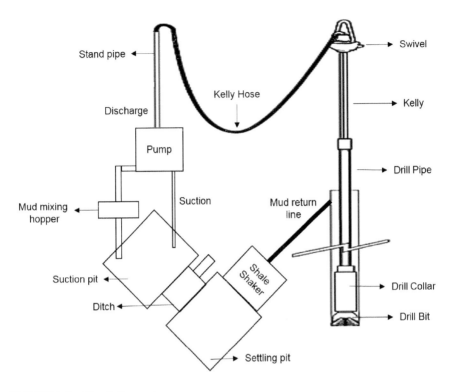

FIGURE 5.4 Circulating system.

Usually, drilling operations are carried out at locations where power is not available. Therefore, diesel-powered internal combustion engines are used to drive the generators to obtain electrical power.

5.2.5 WELL CONTROL AND MONITORING SYSTEMS

To prevent the uncontrolled flow of hydrocarbons to well fluids to the surface, well control systems are employed. When the formation pressure exceeds the hydrostatic pressure exerted by the mud, the formation fluids will begin to enter the well bore. An influx of formation fluid into the wellbore is referred to as a kick. The well control systems are used for detecting the kick, closing in the well, removing the formation fluid, and keeping the well safe for future operations. If the uncontrolled flow is not prevented, it will result in a blowout leading to damage to life, property, environment, and reserves. Primary well control is obtained by maintaining a higher hydrostatic pressure, as compared to the formation pressure. Secondary well control is obtained by the use of valves to withstand the flow of fluids from the well until the well is brought under control. Blowout preventers are used for well control and they seal the well from the top. In the case of land rigs, blowout preventers are maintained below the rig floor, while for floating platforms, blowout preventers are usually placed on the seabed. Annular and ram-type preventers are the most commonly used configurations of the BOP. Continuous monitoring of the drilling operations is essential for safe operations. Most of the problems can be avoided if an early detection of the issue is reported. Several gauges are provided to the driller to obtain the operational parameters. Mud-logging is also carried out to correlate the well properties with the well in the vicinity.

5.3 BLOWOUT

Blowout is generally referred to as an uncontrolled release of well fluids to the surface due to the failure of pressure control systems. Primary and secondary control system prevents the formation of fluid from escaping to the surface. Valves like blowout preventers are used to close the well at the surface. In certain cases, formation fluids can flow into another underground formation at a lower pressure at a different depth. Such a situation is referred to as an underground blowout and is difficult to control.

Blowout can be caused due to various reasons:

a) Reservoir pressure: Due to high reservoir pressure, a sudden release of formation fluids to the surface may lead to blowout without properly rated well control equipment.

b) Formation kick: If the hydrostatic pressure of the drilling fluid falls below the formation pressure, the well fluids shall enter the wellbore and could escalate to a blowout. The situation could even be more severe when there is a rapidly expanding gas also present.

c) Well control: Upon detecting the kick, the well bore is isolated by blowout preventers and closing in the well. The influx of fluids is circulated out in a controlled manner. Quick response to kick and method of response is very crucial for averting a blowout. There are several instances in which a blowout can occur. It can happen during the drilling phase, well testing, completions, production, and also during workover.

5.3.1 SURFACE BLOWOUT

A surface blowout can be catastrophic, causing major damage to the drilling rig. The fluid escaping out can cause the drill string to eject out of the well. The well fluids comprising of oil, produced water, drilling fluid, and natural gas along rocks are spewed into the atmosphere and surroundings. Friction and also sparks caused by the movement of various particles under high velocity can cause the blowout to ignite. The duration of a blowout, depending on the reserves and the reservoir pressure, may take longer to subside (Mayerhofer et al., 2010). Therefore, relief wells are drilled to intersect the well causing blowout and commence well control activities.

5.3.2   Subsea Blowout

Equipment failure and improper balancing of the hydrostatic pressure and formation pressure generally lead to the blowouts in subsea. In the event of a loss of well control, hydraulically actuated blowout preventers are actuated to arrest the flow of hydrocarbons to the surface. To operate in an offshore environment, the maintenance of safety systems is very crucial. Despite the well control equipment being present, the operating companies have to provide an oil spill response plan and also a well containment plan before drilling the well.

5.3.3   Underground Blowout

The movement of high-pressure well fluids from a formation to a low-pressurized zone within the formation. The movement of high-pressure well fluids usually happens from a deeper zone to a shallower zone.

5.4   KICK

A kick is caused when the formation pressure is higher than the hydrostatic pressure exerted by the drilling mud. This shall result in the entry of formation fluids to the wellbore. Various causes are resulting in a kick.

5.4.1   Insufficient Mud Weight

In certain geological conditions, abnormal formation pressures are experienced. Abnormal formation pressure exerts higher pressure as compared to normal conditions. The mud weight exerted during the normal operation will not be sufficient for an abnormal zone and can result in a potential kick leading to a potential blowout.

5.4.2   Improper Hole Fill-up

One of the prominent causes that can result in a kick is improper filling up of the hole during the trip. As the drill string is pulled out, there is a drop in the mud level inside the well bore. Therefore, the level of the mud in the wellbore should be maintained during the tripping operations. Methods like the trip tank method and pump stroke measurement method are used to ensure adequate fill-up of the wellbore.

5.4.3   Swabbing

The retrieval of the drill string from the wellbore creates a negative pressure, also termed as swab pressure. Due to the negative pressure, the effective hydrostatic pressure in the wellbore is reduced and also below the bit. The effect caused can result in the entry of formation fluid into the wellbore.

5.4.4 Cut Mud

It is a rare situation where the mud is contaminated by the gas resulting in the kick. The overall hydrostatic pressure reduces as the gas expands while being circulated to the surface. This reduction in overall hydrostatic pressure could result in an influx of well fluids into the wellbore resulting in a kick.

5.4.5 Lost Circulation

Lost circulation of drilling mud into the formation could also result in a kick. Lost circulation causes a decrease in the hydrostatic pressure due to the short mud column. Kick resulting from lost circulation is severe, as a large volume of well fluid could enter into the wellbore before it is observed at the surface.

Warning signs of the kick include an increase in the torque, and cutting size variation. Mud property variation, change in rate of penetration.

5.5 WELL TYPES

Based on the well construction profile, the wells can be classified as vertical, deviated/directional well, high angle (S shape), horizontal well, extended reach drilling (ERD) well, and multilateral. A vertical well is a traditional technique used to access the subsurface reservoir of hydrocarbons by drilling vertically into the ground. During the drilling operation, the direction and deviation of the wellbore are controlled to reach a predetermined target depth, causing a directional well. High angle well is made up of a vertical section, a kickoff point where the well begins to deviate followed by a build-up section, a tangent section, a drop-off section, and a hold section up till the target, as shown in Figure 5.5. Horizontal well has an inclination of 90 degrees from the vertical while an extended reach drilling (ERD) well is the other name for the directional drilling of long horizontal well. ERD aims at reaching a larger area from a given drilling location and also maintaining a longer distance to maximize productivity and drainage. The multilateral well has multiple branches from one vertical well. It enables a better connection between the well and the formation resulting in higher production. A graphical representation of the wells with various profiles is shown in Figure 5.5, while an S-shaped well is shown in Figure 5.6.

5.6 INSTALLATION AND COMMISSIONING

In a well completion, the actual casing size and tubing size are determined by design criteria from suitable international standards. An example of a well completion procedure is shown in Figure 5.7. To obtain the sequence of well completion, the following steps are to be undertaken.

Installing the 30" Conductor: A truck-mounted pile driver is used to drive the large-diameter pipe to about 100 ft below the ground. The conductor pipe prevents the collapse of unconsolidated rocks present near the surface as the drilling continues

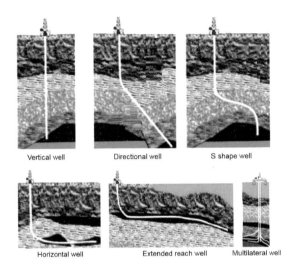

FIGURE 5.5   Types of well.

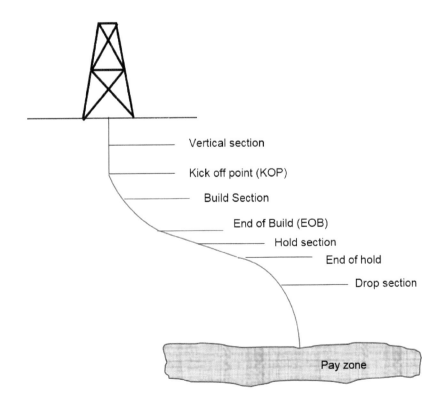

FIGURE 5.6   S-shaped well.

deeper. Once the conductor is installed, the preparation is made for the subsequent operations.

Drilling and Casing the 26" Hole: A drill bit smaller than the diameter of the conductor is used to drill the section. Since the internal diameter of the conductor is approximately 28", a 26" diameter drill bit is used in the current section. Generally, a depth of 2000 ft is drilled in this section covering several unconsolidated formations. Drilling fluid is circulated from the drill pipe to the annulus to bring the cuttings to the surface. The cuttings are separated from the drill fluid and are circulated back to the drill pipe. When a depth of approximately 2000 ft is reached, the drill string is pulled out and the surface casing is run in. The surface casing in the configuration mentioned in the example has an outer diameter of 20". The casing is made up of an incremental length of 40 ft and threaded connections are made at the ends. A slurry of cement is injected into the annular space between the casing and the borehole. It acts as a seal between the borehole and casing and also prevents the caving of the formation into subsequent sections.

Drilling and Casing the 17 ½" Hole: Once the cement sets hard, wellhead housing is attached to the top of the 20" casing. The wellhead housing supports the weight of the subsequent sections. To drill further, a blowout preventer is attached to the top

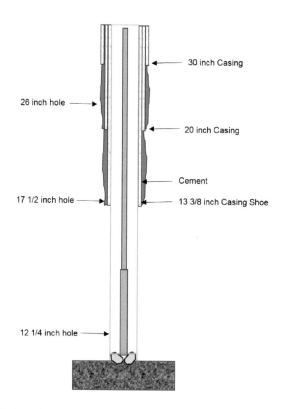

FIGURE 5.7   Casing scheme.

of the casing for well control purposes. Once the blowout preventer is installed and pressure tested, the 17 ½" hole is drilled to 6000 ft. Intermediate casing 13 3/8" is run into the borehole and cemented. Once the cement sets, BOP is nipped down and the wellhead spool is attached to the top of the casing. The wellhead spool shall support the subsequent casing and blowout preventer needed for the successive section.

Drilling and Casing the 12 1/4" Hole: Blowout preventer is tested and re-installed to drill the 12 ¼" hole through the pay zone. The cuttings obtained are analyzed for a trace of oil present in them. The presence of gas will be detected by the gas detectors placed on the flow line. In case oil or gas is present, the formation is evaluated more thoroughly. The drill string is retrieved and the wireline logging tools are used to obtain the petrophysical logs. Coring operations will also be carried out to obtain a large cylindrical sample from the section. A drill stem test is carried out to determine the production capabilities of the drilled well. Once the results from the logging and drill stem test are positive the company decides to complete the well. A production casing 9 5/8" is run into the well and cemented. A tubing of 4 1/5" outer diameter is run into the casing string. A packer is used to seal off the annulus between the production casing and the tubing. Depending on the application, the packer can be retrievable or permanent. The packer is locked in its position by mechanical means or by use of hydraulic force.

5.7   PRODUCTION PROCESS

Offshore production engineering forms an essential component of the operations, as the return on investment depends on the quantity of hydrocarbons produced by the well. Once the well is drilled and completed, the terminology of production covers all activities carried out to retrieve the crude oil from the well bore to the processing facility. The produced crude is processed and transported finally to the oil and gas refinery. To improve the performance of the well, it may be subjected to workover operations. Over some time, with a decrease in reservoir pressure, the well may require artificial lift support. The processing of oil and gas onshore and offshore is done similarly. However, the processing and injection facilities need to be accommodated in an integrated manner to manage the space constraints and also meet the operational requirements.

5.7.1   HORIZONTAL SEPARATOR

The initial phase of gross separation between liquid and vapor takes place as the fluid enters the separator through the inlet, where its momentum is disrupted by the inlet diverter. Gravity causes liquid droplets to separate from the gas stream, collecting at the separator's bottom. Over time, gas evolves from the oil and rises to the vapor space, reaching "equilibrium" due to sufficient retention time from the liquid collection. In the event of occasional liquid slug formation, it also provides a surge of volume. Liquid exits through the controlled liquid dump valve, operated by a level controller that senses changes in the liquid and adjusts the dump valve accordingly.

Gas and oil mist flow over the inlet diverter and move horizontally through the gravity settling section positioned above the liquid. Despite not being separated by the inlet diverter initially, small liquid droplets undergo separation in this section due to gravity. Due to their small diameter, certain drops evade separation in the gravity-settling section, necessitating their passage through a coalescing section or mist extractor. Utilizing vanes, wire mesh, or plates, this section offers ample surface area for coalescence, effectively eliminating these minute liquid droplets as the last step of separation before the gas exits the separator.

A pressure controller is installed on the gas outlet to regulate the pressure within the separator. This controller also detects pressure variations within the separator and transmits signals to adjust the opening and closing of the pressure valve accordingly. By controlling the rate of gas exiting the vapor space, the separator's pressure is maintained as shown in Figure 5.8. Typically, horizontal separators operate at half liquid capacity to optimize the gas-liquid interface's surface area. Due to their smaller size, horizontal separators are generally less expensive compared to vertical separators for a given gas and liquid flow rate. They are frequently employed in flow streams characterized by high gas-liquid ratios and foaming crude.

5.7.2 VERTICAL SEPARATOR

In a vertical separator, the vessel receives the inlet flow from its side, and initial separation occurs through the inlet diverter, akin to a horizontal separator. Liquid then descends to the collection section, usually without any internal components aside from a still well for level control, as shown in Figure 5.9. This still well,

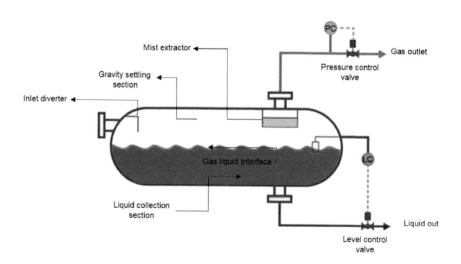

FIGURE 5.8   Horizontal separator.

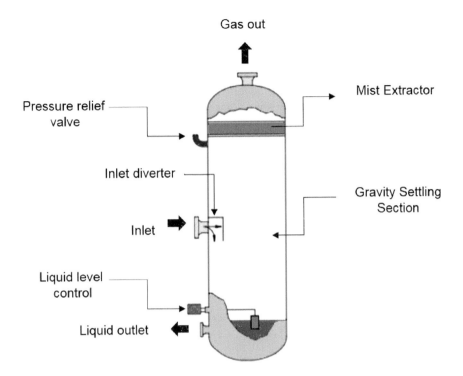

FIGURE 5.9   Vertical separator.

often a simple open box or tube, serves to stabilize the level control by preventing interference from waves. Liquid continues downward to the outlet. As equilibrium is established, gas bubbles move against the liquid flow, eventually reaching the vapor space. Similar to a horizontal separator, the level controller and liquid dump valve functions.

The gas travels vertically over the inlet diverter and toward the gas outlet. Secondary separation takes place in the upper gravity settling section, where liquid droplets fall vertically downwards opposite to the upward gas flow. The settling velocity of a liquid droplet is directly related to its size. A droplet is carried upwards and out with the vapor if the diameter is too small. Small liquid droplets are captured by a mist extractor before they can exit the separator. Pressure and level are regulated like that of a horizontal separator. Commonly used in flow streams with low to intermediate gas-liquid ratios, vertical separators are used well for the production containing sand and other sediments; hence, used with a fake bottom to handle sand production.

5.7.3   SPHERICAL SEPARATOR

This separator follows a similar structure with four sections, as shown in Figure 5.10. Spherical separators, a specialized type of vertical separator, lack a cylindrical shell

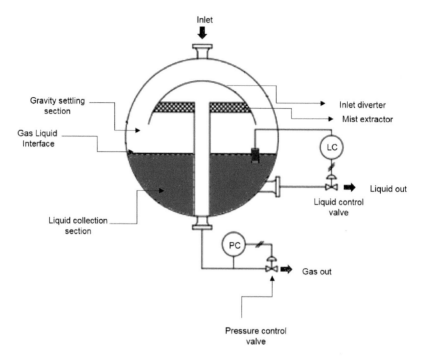

FIGURE 5.10   Spherical separator.

between their heads. Fluid enters from the top through the inlet diverter, dividing into two streams. Liquid descends to the collection area via openings in a horizontal plate slightly below the gas-liquid interface. The thin liquid layer facilitates gas separation, allowing gases to rise to the gravity-settling section. Gases exiting the liquid pass through the mist extractor before leaving through the gas outlet. Liquid level is regulated by a float connected to a dump valve, and pressure is controlled by a back pressure valve.

Originally designed to theoretically take the best characteristics of both horizontal and vertical separators, spherical separators have, in practice, encountered the worst characteristics and are exceedingly challenging to size and operate. While they may be highly efficient from a pressure containment standpoint, their limited liquid surge capability and fabrication difficulties mean they are seldom utilized in oil field facilities. For this reason, further discussion of spherical separators will not be pursued.

5.8   SECTIONS OF SEPARATOR

5.8.1   INLET DIVERTER SECTION

The fluid that enters the vessel typically consists of a high-velocity turbulent mixture of gas and oil. Owing to this high velocity, the fluids carry significant momentum into the separator, which is abruptly altered by the change in the flow of direction by absorption of the liquid's momentum, facilitating the separation of liquid and gas by the inlet diverter, also known as the primary separation sector. This process initiates the initial "gross" separation of the two phases.

5.8.2 LIQUID COLLECTION SECTION

Situated at the bottom of the vessel, the liquid collection section offers the necessary retention time for any entrained gas in the liquid to dissipate into the gravity settling section. It also offers surge volume to manage intermittent slugs. The level of separation depends on the provided retention time, which is influenced by the separator's liquid capacity, the fluid entry rate, and the density difference between fluids. Liquid-liquid separation typically demands longer retention times compared to gas-liquid separation.

5.8.3 GRAVITY SETTLING SECTION

Upon entering the gravity settling section, the gas stream experiences a decrease in velocity, allowing small liquid droplets previously entrained in the gas and not separated by the inlet diverter to separate due to gravity and descend to the gas-liquid interface. The sizing of the gravity settling section ensures that liquid droplets larger than 100 to 140 microns settle to the gas-liquid interface, while smaller droplets remain suspended with the gas. Liquid droplets exceeding 100 to 140 microns are deemed undesirable as they can overwhelm the mist extractor at the separator outlet.

5.8.4 MIST EXTRACTOR SECTION

The gas exiting the gravity settling section typically carries small liquid droplets, usually less than 100 to 140 microns in size. Before exiting the vessel, the gas passes through a coalescing section or mist extractor, where this section employs coalescing elements that offer substantial surface area for coalescence, effectively removing the small liquid droplets. As the gas traverses the coalescing elements, it undergoes multiple directional shifts, where the liquid droplets are unable to swiftly adjust to these rapid changes in flow direction due to their greater mass. Therefore, these droplets collide with and accumulate on the coalescing elements, eventually descending to the liquid collection section.

5.9 HEATER TREATER

The heater-treater offers an improvement over the gun barrel and heater system and comes in various designs to accommodate different conditions such as viscosity, oil gravity, and flow rates. Compared to gun barrels, heater-treaters are cheaper upfront, easier to install, more efficient at heating, and offer greater flexibility. However, they are more complex, have less space for sediment storage, and are more sensitive to chemicals. Due to their smaller size, heater-treater has shorter retention times (usually 10 to 30 minutes) compared to other vessels. Internal corrosion in the down-comer pipe is common, and sediment buildup on the walls or bottom can cause issues like elevated interface levels and liquid carryover. Regular inspections are needed to check for corrosion, sediment build-up, and scale accumulation.

5.9.1 HORIZONTAL HEATER-TREATER

For most multi-well flow streams, horizontal heater-treaters are often necessary. These units are usually comprised of three main sections:

1. **Front Section:** At this initial stage, the entry of fluid into the heater-treater takes place. Here, a water-washing process occurs, separating impurities.
2. **Oil Surge Chamber:** As the fluid progresses upward, it encounters heating elements, further separating components. Free water continues to separate, and precise water level control is maintained to prevent operational issues.
3. **Coalescing Section:** In this phase, water droplets are effectively separated from the oil. Clean oil is collected, while residual water droplets descend.

Fluids enter the front section through a fluid inlet and flow down over the deflector hood, separating heavier materials (water and solids) at the bottom and lighter materials (gas and oil) at the top, as shown in Figure 5.11. Free gas is removed through the gas equalizer loop, while oil and emulsion are heated as they rise past the fire tubes and are collected in the oil surge chamber. A level controller in the oil surge section operates the dump valve on the clean oil outlet line to regulate oil flow. The treated

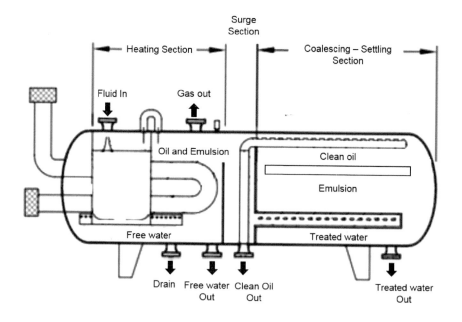

FIGURE 5.11 Horizontal heater-treater.

oil rises to the clean oil collector while coalescing water droplets fall counter-current to the rising oil continuous phase. The front section must be sized to handle the settling of free water and heating of the oil, while the coalescing section must provide adequate retention time for coalescence and allow water droplets to settle downward counter-current to the oil flow.

Nowadays, most horizontal heater-treaters do not use fire tubes, and heat is added to the emulsion through a heat exchanger before it enters the treater. Some heater-treaters are designed with only the coalescing section, with the inlet pumped through a heat exchanger to a treater operating at a high enough pressure to prevent gas evolution in the coalescing section.

5.10   OFFSHORE PRODUCTION SYSTEM

The production system can be fixed or floating in an offshore location depending on the configuration of the platform. In a fixed platform the production is carried out with the support of conductor pipes extending from the seabed to the platform. In the case of a floating platform, the oil and gas from the subsea are transferred to the surface facility by marine risers (Bai, 2001).

5.10.1   FIXED PRODUCTION PLATFORM

This system includes fixed production platforms and also the offshore terminals like SBM and so on. The fixed production platforms can be divided into three types: the well platform, the process platform, and the injection platform.

Well platforms are usually four-legged jacket platforms having four to 12 slots of conductor extending from the seabed. These conductors are enclosed within the frame of the jacket and have deck-completed oil and gas wells that produce well fluids from the reservoir beneath. These well fluids are received at the wellhead on the deck of the platform. The produced well fluids are gathered at the header from various wellheads and transferred through an export riser to the processing platform through a subsea flow line. Well, platforms consist of primarily three decks of which two are the main deck and cellar deck. The cellar deck consists of the Christmas tree. The third deck, which is the helideck, is used for transportation of operating personnel. Cranes and hoists are also provided for mechanical operations. These platforms are usually unmanned but have a boat landing facility with a bunkhouse in case of any emergency operations.

Safety valves such as emergency shutdown valves, surface-controlled subsurface safety valves, and high–low valves are integrated into the system to ensure safety during operations. These valves are operated through telemetry with the support of the command center in the process platform. As carried out in an onshore location, the well streams from various conductors are connected to headers. The header where the majority of wells are connected is referred to as the group header, while the test header connects only to one of the wells. The test header directs the well stream to a test separator, which is a three-phase separator. The performance of the well in terms of water cut, production rate, GOR, etc., is obtained. Depending on the predefined schedule, wells are connected to the test header.

For powering the platform, generators are used with auxiliary support from the solar panels. Fire water pumps are installed for emergencies and also chemical injection pumps are provided for injection of chemicals. In case of any leak of gas or fire, the detection system gets activated and the platform is shut down. In case of any emergency during the operations, life jackets, buoys, and rafts are provided on the platform.

5.10.2 PROCESS PLATFORM

It can be compared to that of a group gathering station at an onshore location. The process platforms are connected to a network of well platforms. It consists of various pieces of equipment for the separation, treatment, and processing of the well fluids received from various well platforms. Once the well fluid is processed in the platform and separated into oil water and gas, they are metered and transported to the refinery using pumps and subsea pipelines. The process platform houses a command center that regulates the entire process going on in the platform. One of the major works of the process platform is to separate the oil, water, and gas from the well fluids. The produced water is treated to industrial standards and is disposed of. Gas is conditioned to power up the processing facility and other auxiliary equipment. The excess gas is flared off from the platform. First, the well stream is received at the manifold and moves through high- and low-pressure inlet separators. Separators can be horizontal or vertical, depending on the space available, the presence of gas, and the production rate. The crude oil moves to the surge tank from the separators. The produced water is conditioned using corrugated plate interceptors and is subjected to further processing, depending on the amount of oil content present in it. Once the produced water meets the industrial norms, it is disposed of into the sea.

The gas obtained from the separator is sent to the scrubbing unit. Further, the gas is dehydrated and compressed to be dispatched to the onshore refinery. Excess gas is scrubbed off the liquids and flared through a flaring system connected to the platform using a tripod bridge. The power generation is primarily by generators powered by gas or diesel. A generator driven by diesel is installed for emergency power. The gas obtained from the high-pressure separator is conditioned before supplying it as fuel for the generator. A diesel fuel system provides fuel to the cranes, fire water pumps, and generators. The process platform also consists of a jet fuel system, which is used to provide aviation fuel to the helicopters. The air system consists of compressors that are used to provide dry air to support pneumatic valves and instrumentation. For general-purpose operations, non-dry air is also provided by the air system. To provide portable water, a reverse-osmosis operation is carried out in the portable water system and is collected in the storage tanks. To avoid algal growth in subsea pumps, electrolysis of seawater is carried out to obtain sodium hypochlorite and is distributed to the pumps. Cranes and hoists are used for moving materials on the platform and also for supporting maintenance operations. A vent system is used to collect low-pressure gas obtained from different equipment. The collected gas is scrubbed and liberated to the atmosphere. Flame arrestors are used to avoid any flashing of the liberated gases. A waste-heat-recovery system uses a heat exchanger to recover the heat from the exhaust gases liberated from the generators. The recovered heat is utilized in various useful applications in the platform. Various low and high-pressure gases released from various equipment are flared at the flare tip.

Various communication channels are used for communicating with other platforms, vessel communication, offices located onshore, and emergency distress. Operational commands with the platform are usually carried out using a walkie-talkie. Nondirectional beacons and aero-trans-receivers are used to assist the helicopter operations. Offshore support vessels are utilized for the transit of personnel and materials through the sea. These vessels also play a major role in evacuation during emergencies. Multi-support vessels are utilized for inspection and maintenance activities related to equipment and pipelines.

5.10.3 INJECTION PLATFORM

It requires the processing of water for injection and also requires a good platform to carry out the injection activities. In the case of the injection water process platform, the processing facility is supported by lift pumps that enable the lifting of seawater and also chlorinators that electrolyze the seawater to produce sodium hypochlorite. The obtained sea water is subjected to two layers of the filtration process—i.e., coarse and fine. The number of coarse and fine filters is maintained, depending on the amount of particulate matter present in the water. Once the filtering is completed in the coarse stage, water is pumped into the fine-filtration level. Beyond a certain threshold of differential pressure in the filter, a backwash is carried out. From the filter, water is moved to the de-oxygenation tower, where an oxygen scavenger is added with the support of spray nozzles. Booster pumps are used to increase the pressure before the water is moved to different injection platforms through injection lines.

In the case of water injection well platforms, they are solely meant for injection of water and are similar to the well platforms. The process platform delivers the treated water and is transferred to the injection well after metering. The platform has a shutdown panel that is driven by a hydraulic power package. A water injection manifold skid is used to distribute the incoming treated water from the process platform to multiple wells.

5.11   PRODUCTION FACILITY IN A FLOATING PLATFORM

An increase in energy demand and potential hydrocarbon prospects led to the development of several forms of floating oil and gas platforms for tapping oil and gas from deep waters. The advancements in subsea technologies contributed to it even further. These platforms could be used in ultra-deepwater environments and could be reused. The top sides of these platforms house the necessary production operations to be carried out. Marine risers serve as an interface between the subsea facility and the topsides.

5.11.1   TOPSIDES

The topside consists of several modules for carrying out production operations. The most prominent system modules are process, utilities, and support and safety. To take into account the safety of the personnel on board, the living quarters and flare system are placed on the windward and leeward sides, respectively. The fuel sources are separated from the ignition sources to the maximum possible extent. The equipment is placed in an optimized manner to reduce the piping. The water handling equipment and the utilities are placed near the living quarters.

5.11.2 PROCESS SYSTEM MODULE

It is comprised of three modules—i.e. oil, produced water, and gas. Auxiliary modules like venting and gas flaring modules, main oil line pump modules (MOL), and process gas compressor modules will also assist in the operations. In the oil module, oil is separated through a high-pressure separator followed by a low-pressure separator. The oil further moves to the surge tank, which is operated depending on the mode of delivery of the crude oil. The surge tank operates in pressurized mode when the oil has to be sent via pipelines. This enables the lighter fractions to remain in the oil and avoid being flared. The surge tank is operated at atmospheric pressure when the oil is to be delivered in tankers. The oil in tankers needs to be stabilized, and the lighter fractions are knocked off. The stabilized crude from the floating production unit is transferred to the shuttle tankers. To determine the quantity of oil dispatched, the flow rate integrator is maintained in the centralized control panel. Further, an electrostatic heater and chemical treatment may have to be carried out in the case of emulsified crude. The optimization of the chemicals used has to be determined under laboratory conditions.

In the gas module, the compression system receives the high-pressure gas through the inlet manifold. The high-pressure gas is further dehydrated in the dehydration module. In the case of sour gas, a gas sweetening module needs to be additionally integrated. This high-pressure gas is used as a source of fuel for operating the turbines and compressors. Also, the high-pressure gas is used in the gas lift type of artificial lift technique or even reinjected back into the reservoir. The gas may also be dispatched to a nearby consumer, which is quite rare at remote offshore locations. An orifice meter is used to determine the gas flow rate from the separator. Also, the gas flow integrator present in the control panel measures the quantity of gas dispatched. Low-pressure gases from the separator and surge tanks are flared. In the produced water module, produced water obtained from the separators and tanks is passed to the conditioner, where the oil is skimmed and treated further. The treated water is disposed into the sea by complying with regulatory requirements or re-injected back into the reservoir.

5.12 FLOATING PRODUCTION STORAGE AND OFFLOADING (FPSO)

A FPSO unit is a floating vessel used for hydrocarbon processing and storage at an offshore location. Hydrocarbons are received in an FPSO by a subsea production facility or a nearby production platform. The hydrocarbon received is processed onboard and is offloaded to tankers. FPSOs are used in remote offshore locations where it may not be economical to use pipelines to transfer the processed crude to the offshore facility. FPSOs are ship-shaped, and therefore, from a distance, they cannot be distinguished from oil tankers. Some of the FSPOs are built while the others are converted oil tankers. The produced gas from the reservoir is separated in the FPSO but is generally not stored. The produced gas is re-injected back to the reservoir or is supplied to a nearby consumer if any. The turnaround time of an FPSO depends on the size of the FPSO. The topsides of the FPSO should accommodate the equipment for processing, accommodation, and the required utilities. The ballast capacity should accommodate the effect of motion. It should also accommodate the turret system.

To operate a FPSO in extreme offshore environments, the mooring systems have the capability to weathervane. Mooring lines are attached to the turret, which enables the FPSO to rotate and orient along the direction of the wind; the turret may be placed inside the hull or at the bow or stern. The axis of the rotation of the FPSO is about the turret, irrespective of its position. The turret serves as an interface between the subsea production system and the FPSO. The production, export, and gas reinjection risers are connected to the turret. Apart from the risers, other auxiliary lines like electrical, hydraulic, chemical, and pneumatic are connected to the turret.

A turret consists of a swivel stack that directs the produced well fluids to topsides through a network of flow lines. In a few of the FPSO designs, under extreme weather conditions, the turret can be disengaged, with the risers and mooring connected to it to a predetermined depth. The FPSO can leave the location and continue production at a later date. FPSO has to off-load depending on the storage capacity and the turnaround time of the shuttle tankers has to be planned accordingly. FPSO units have storage tanks of high capacity, and some designs can accommodate up to 2 million barrels. FPSOs are effective in remote deep-water locations where it is not economical to lay pipelines. The concept avoids the need for investment in long-distance pipelines. It is an attractive solution for marginal oil fields which shall be exhausted in a few years (Cairns, 1992). Once the reservoir is depleted, the FPSO can move to a newer location. This initial cost of the FPSO is high, but its maintenance cost is limited.

5.13 OFFLOADING TERMINALS

It would be advantageous to transfer the produced crude using pipelines. In case pipelines are absent between the offshore production facility and the refinery, the only option left is to use shuttle tankers or export tankers. The shuttle tankers are loaded and unloaded in an offshore terminal. Offshore terminals are currently being used for transporting petroleum products and liquid freight. Over the years, the capacity of the cargo tanks has increased several folds to cater to the growing volume demands of the offshore industry. Currently, larger tanks are being used to transfer crude oil from offshore locations, which is also advantageous, considering the turnaround time. Currently, there are several configurations of offshore terminals, like single-point mooring, multi-buoy mooring, and also sea islands.

5.14 SINGLE-POINT MOORING

5.14.1 Catenary Anchor Leg Mooring (CALM)

The buoy is anchored to the seabed using chains, as shown in Figure 5.12. The buoy is circular in shape and is watertight. Along the external surface of the body of the buoy, a circular skirt is welded where the anchor chains are connected. A turntable is

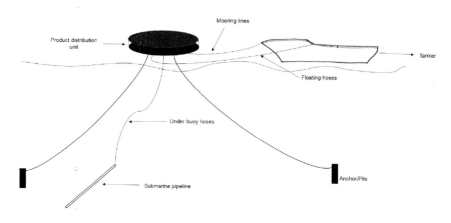

FIGURE 5.12   CALM buoy.

located on the top of the buoy that can freely rotate about its vertical axis. Mooring is attached with one point on the tanker and the other on the turntable, which enables the tanker to swing freely. The turntable also houses the oil handling system. Floating hoses are used to connect the manifold of the tanker and the center swivel present on the buoy. The center swivel is attached to the subsea pipelines using underwater hoses. The installation can be completed in a short period, but the maintenance cost is high due to the wear of hoses and chains. The components are to be replaced in case of any deviation observed during the periodical inspection.

5.14.2   Single Anchor Leg Mooring

The SALM system consists of a tubular under-buoy pipe connected to the buoyancy chamber on the top. The under-buoy pipe acts like a riser and transfers liquid from the tanker side to the subsea pipeline, as shown in Figure 5.13. The floating buoy is connected to the buoyancy chamber using an anchor chain through the anchor swivel and the fluid swivel housing. Mooring lines are attached from the tanker to the floating buoy. The floating loading hose is attached to the fluid swivel housing through which the produced crude or liquid is transferred through the under-buoy pipe. The under-buoy pipe is connected to the subsea pipeline using a universal joint. The assembly is housed on a base which is piled to the seabed.

One of the major advantages of SALM is that the tanker can approach closer to the facility, as the buoyancy chamber and buoy pipe are under water, which could be placed below the tanker keel. Also, since the floating buoy is vertically moored to the anchor swivel, the tanker can anchor itself to the seabed without the risk of entanglement. Since the swivel unit is underwater, the overall wear of the hose is reduced. The system requires frequent inspection due to its complexity; also, the repair would involve complete removal of the system from the seabed and reinstallation causing a downtime in operations. Since the floating buoy is attached to a chain, the wear of it would be even higher.

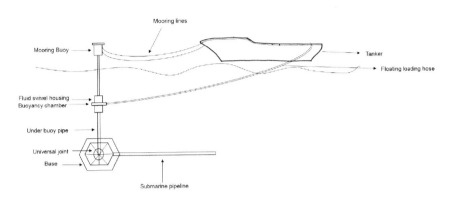

FIGURE 5.13   SALM buoy.

5.15   SINGLE-POINT MOORING TOWER (SPMT)

5.15.1   FIXED TOWER MOORING

A tower mounted with a turntable is attached to the seabed. The point of mooring and the swivel is located on the turntable. A steel riser pipe acts as an interface between the manifold on the turntable and the subsea pipeline. This arrangement of having a riser steel pipe omits the use of floating hoses. Since underwater swivels and chains are not being used, the maintenance cost is lower. A higher transfer rate of crude oil is achieved due to the replacement of underwater hoses with riser pipes. Sacrificial anodes are used to safeguard the tower from corrosion which can make it operational for over three decades. However, overriding the oil tanker onto the tower can cause serious damage to the structure as well as the tanker. The cost of repair will be significantly higher and will take several man hours for completion.

5.16   TERMINATION OF WELLS

Once the oil and gas well has reached the end of its life, it is decided to plug and abandon it. Such operations are carried out due to depleted reservoirs, subsidence-induced problems, lack of well integrity, etc. Such operations may also have to be carried out in offshore locations in case of an accident between the platform and support vessels (Khalifeh and Saasen, 2020). A well may also be plugged when it may not be economical to operate compared to the quantity of crude produced. It is also essential to terminate a well because a possible pressure builds up in the future leading to a flow of hazardous fluids into areas like groundwater, the atmosphere, and also the marine environment. Suitable equipment will be used to create a permanent barrier across the formation to prevenient any untoward incidents in the future as shown in Figure 5.14.

Once the production activity of the well is stopped, the well status needs to be defined. There are three possible statuses for the well in this condition, i.e., suspended,

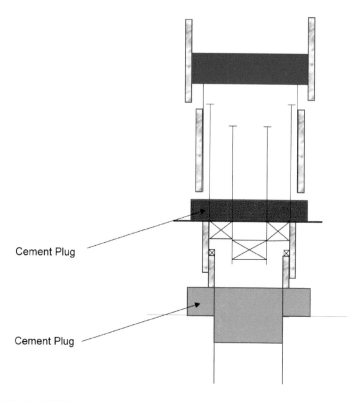

Cement Plug

Cement Plug

FIGURE 5.14　Well Barrier Envelope

temporarily abandoned, and permanently abandoned. The status of 'suspended' is provided when during construction or intervention operation. However, during the operation, the well control equipment was present on the well. The status of 'temporarily abandoned' is provided when the well control equipment is removed. Re-entry or permanent abandonment may be thought of at a later stage. Status of 'permanent abandonment' is provided when the well or a part of it is plugged permanently with no intention of re-entry.

5.17　WELL BARRIER ENVELOPE

5.17.1　Primary and Secondary Well Barriers

The term "barrier" refers to an obstacle that prevents the movement from one place to another. About plug and abandonment, a barrier is an obstacle that prevents the release of well fluids in an uncontrolled manner. In the two-barrier approach, there is a primary barrier and a secondary barrier that work independently. Both primary and secondary barriers prevent the flow from a possible source of inflow. The secondary barrier is used for redundancy in case the primary barrier fails.

5.18   ENVIRONMENTAL PLUG

Apart from the primary and secondary plugs, there is an additional plug that is installed near the surface called the environmental plug. It safeguards the open annulus from the outer environment. It is also sometimes referred to as an open-hole plug or surface plug. As the environmental plugs cannot withstand high pressure, barriers are deployed. As the casings near the seabed are cut and retrieved, the environmental plug seals the open annulus left behind. The plug also prevents the swabbing of seawater through the annuli into the formation and minimizes the chances of leaking from any potential source near the surface.

5.19   PHASES OF WELL ABANDONMENT

Different phases of well abandonment are explained as follows:

5.19.1   PHASE 1: RESERVOIR ABANDONMENT

In the phase of reservoir abandonment, the wireline unit is rigged up after inspection of the wellhead. A caliper log is run by the wireline unit to evaluate well access and also the health of the production tubing. This initial phase of operation and its results play a significant role in the overall execution plug and abandonment operation. To evaluate the well integrity, an injection test is carried out. Once it is ascertained that the well integrity is safe for further operations, bull heading of cement slurry is carried out to plug the formation. Pressure testing is carried out to determine the strength and quality of the cement plug formed. Since the well integrity was within acceptable limits, the operations were carried out using a wireline unit and were less. However, when the well integrity is not satisfactory for further operations, a work-over rig is stationed along with a blowout preventer nipped up. As the primary and secondary barrier is secured the reservoir abandonment phase is completed.

5.19.2   PHASE 2: INTERMEDIATE ABANDONMENT

In the phase of intermediate abandonment, barriers are set to separate the intermediate hydrocarbon-bearing zones. The phase also involves milling operations and retrieval of casings. In case the production tubing is not retrieved in phase 1, the tubing could be partly recovered. Once all the possible potential flow paths are secured, the intermediate abandonment phase is completed.

5.19.3   PHASE 3: REMOVAL OF CONDUCTOR AND WELLHEAD

The wellhead and the conductor are retrieved after cutting them below the surface or from the level of the seabed. In this phase, the conductor and wellhead are cut below the surface or seabed. These operations are carried out to prevent any incident during regular marine operations.

5.20 ENHANCED OIL RECOVERY

Hydrocarbon recovery is conducted by primary, secondary, and tertiary methods. Primary recovery utilizes natural pressure to bring crude oil to the surface. This recovery procedure involves artificial lift techniques such as sucker rod pump (SRP), gas lift (GL), electrical submersible pumps (ESP), and progressive cavity pumps (PCP) to assist the reservoir. During secondary recovery, the injection of pressurized fluids like water and gas into the reservoir takes place intending to displace the remaining oil and gas left after primary recovery. Tertiary recovery, also known as enhanced oil recovery (EOR), involves the use of different techniques to boost crude oil production, as shown in Figure 5.15. The procedure involves the injection of materials that are typically not native to the reservoir.

"EOR" refers to a collection of methods employed to extract more oil from an oilfield that has completed primary and secondary recovery phases. Usually, the initial phases of oil recovery only manage to extract approximately 30-40% of the original oil in place (OOIP), resulting in a substantial amount of oil still trapped beneath. EOR methods focus on extracting additional oil from reservoirs to raise recovery rates to 60% or beyond, thus prolonging the oilfield's productive lifespan. An overall recovery under various modes is shown in Figure 5.16.

A significant quantity of oil remains untapped, even after employing primary and secondary recovery techniques. Given the escalating energy demand, it becomes imperative to extract the remaining oil. The feasibility of drilling new infield wells might be constrained by the high costs associated with drilling operations. The three primary categories of EOR approaches are: thermal recovery, gas injection, and chemical flooding.

The thermal recovery approach entails sending heat into the reservoir to reduce the oil's viscosity, facilitating its flow. Typical methods include steam injection, which involves the injection of hot steam into the reservoir using wells, and in-situ combustion, which involves the injection of air into the reservoir to start a controlled subsurface fire that heats the oil. The gas injection approach entails injecting various gases into the reservoir to enhance the flow of oil towards the producing wells. Various gases

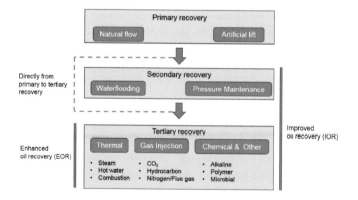

FIGURE 5.15 Modes of recovery in oil and gas reservoirs.

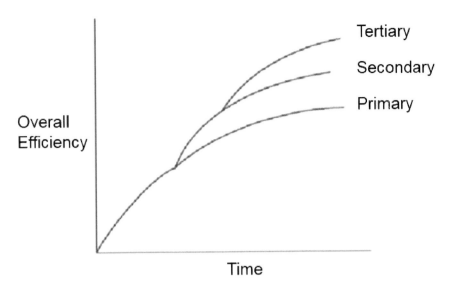

FIGURE 5.16 Overall recovery with time under various modes.

have distinct impacts. Miscible flooding involves injecting hydrocarbon gas. Propane or methane, which are lighter hydrocarbons, blend with oil, decreasing its viscosity and causing it to swell, facilitating its movement. Carbon dioxide (CO_2) flooding involves the dissolution of CO_2 in oil to decrease viscosity and enhance mobility. Injecting CO_2 can capture and store it underground, reducing its environmental impact. Nitrogen (N_2) injection involves using inert N_2 to sustain reservoir pressure and displace oil toward producing wells without blending with the oil. The chemical flooding approach entails injecting chemicals into the reservoir to modify the characteristics of the rock or the oil to enhance the oil recovery. This approach involves techniques like polymer flooding, wherein polymers are added to water injection to enhance its viscosity to improve its ability to sweep oil toward producing wells, and surfactant flooding, wherein surfactants are injected to reduce the interfacial tension (IFT) between oil and water to facilitate the displacement oil by water.

Reservoir characteristics like oil viscosity, temperature, pressure, and rock permeability impact the effectiveness of various EOR procedures. The expenses associated with adopting and sustaining EOR methods must be compared against the possible rise in oil output. Certain EOR methods, such as CO_2 flooding, can offer environmental advantages, but others may necessitate thorough evaluation and mitigation plans. EOR is essential for optimizing oil output from existing fields, which helps ensure global energy security and prolongs the use of important resources.

5.21 EOR IN OFFSHORE ENVIRONMENT

EOR is important in traditional onshore oilfields but encounters distinct problems and opportunities when applied in offshore settings. Offshore operations are intrinsically more convoluted and costly compared to onshore operations. Establishing and up-keeping EOR infrastructure is notably more difficult and expensive when located at sea. Environmental risks are higher at sea due to spills and leaks, necessitating more stringent rules and increased safety precautions while implementing EOR. Space constraints on offshore platforms pose challenges for installing and operating EOR equipment, in comparison to land-based operations. Offshore reservoirs may have heavier and more viscous oil, which could make them more amenable to EOR methods that enhance flow, potentially resulting in better recovery rates than onshore sites. Advancements in subsea technologies, remote monitoring, and automation are enhancing the feasibility and efficiency of EOR operations in offshore locations. Maturing offshore assets are reaching the latter stages of production, making EOR crucial for prolonging their economic viability and optimizing resource efficiency. Thermal recovery by steam flooding is not frequently used offshore because of the energy losses that occur during transportation through long subsea pipelines. Managing subsurface fires during in-situ combustion in a subsea setting is quite challenging.

CO_2 injection is a viable choice for offshore applications. CO_2 that is captured can be transferred and stored in depleted offshore reservoirs, providing environmental and economic advantages. N_2 injection can be used for maintaining pressure and displacing oil. Chemical flooding—specifically, polymer flooding—is less efficient in high-salinity offshore conditions because of the interactions between polymers and salts. Surfactants can be customized for certain offshore environments and have demonstrated positive outcomes in enhancing oil recovery. Despite the enormous constraints of EOR in offshore environments, technological developments and the possibility for increased rates of recovery are driving the ongoing development and implementation of these techniques. Moreover, it is crucial to mention that choosing the best method for EOR in offshore areas involves a comprehensive assessment of technical feasibility, economic feasibility, and environmental consequences. Continuous research and development are essential for improving current EOR methods and investigating new strategies tailored to the distinct problems of offshore operations.

5.21.1 PREREQUISITE KNOWLEDGE

Macroscopic Displacement Efficiency is the ratio of oil volume in contact with displacing agents to the original oil in place (OOIP). Microscopic Displacement Efficiency is the ratio of oil pushed or mobilized by the displacing agents to the volume of oil contacted by the displacing agents. Mobility is the ratio of effective permeability to the phase viscosity.

$$\lambda_w = \frac{k_w}{\mu_w} \tag{5.2}$$

Mobility Ratio is the ratio of the mobility of the displacing phase to the mobility of the displaced phase.

$$M_{w-0} = \frac{\lambda_w}{\lambda_w} = \frac{k_w}{\mu_w} \frac{\mu_0}{k_0} \tag{5.3}$$

Capillary number is the ratio of viscous drag forces to surface tension forces and is given by:

$$C_a = \frac{\mu U}{\sigma} \tag{5.4}$$

Overall Recovery Factor (R_F) is defined for an oil recovery process through secondary or tertiary mode. It is the product of displacement (E_D), areal sweep (E_A), and vertical sweep (E_V) efficiencies.

$$R_F = E_D * E_A * E_V \tag{5.5}$$

Displacement efficiency (E_D) is determined by the reservoir fluid's saturation and mobility. Areal sweep efficiency (E_A) is determined by the areal heterogeneity, pattern type, fluid mobility, and injected fluid volume. Vertical sweep efficiency (E_V) is determined by vertical heterogeneity, fluid mobility, degree of gravity segregation, and injected fluid volume.

5.22 THERMAL RECOVERY

5.22.1 STEAM INJECTION

Steam provides the necessary heat to raise the reservoir temperature and initiate combustion for displacing oil in the thermal recovery process. The steam generator produces the steam that acts as the heat transfer medium. The injected steam transfers the heat to the reservoir formation and the fluids within it. Heat reduces the oil's thickness and enhances its flow. Steam injection for enhancing oil recovery can be classified into two: cyclic steam injection and steam drive or steam flooding.

5.22.2 CYCLIC STEAM STIMULATION

This method involves using the same well for both injecting steam and producing oil. Initially, steam is delivered for a duration ranging from a few weeks to a few months. The steam provided heats the oil within the vicinity of the injection well quickly through convective heating, reducing its viscosity. Steam injection ceases once the desired viscosity is achieved to provide uniform heat distribution in the formation. This helps optimize the oil recovery efficiency at this stage. Subsequently, the well can be operated until the temperature decreases and the oil's viscosity rises again. This marks the commencement of a new steam injection cycle to raise the temperature of the reservoir.

5.22.3 STEAM FLOODING

The injection and production wells in steam flooding have distinct characteristics. Steam is injected into the wells to displace oil, physically pushing it and heating it, lowering its viscosity. Steam is continuously injected into the injection well in the process of steam drive to create and advance a front. Steam floods need continuity to enable steam to push oil toward the production wells. The process is more expensive than cyclic steaming due to the higher steam consumption involved in the operation. However, the technique typically aids in the recovery of a greater amount of the oil. In certain fields with ideal connectivity, both cyclic stimulation and steam flooding approaches can be utilized.

Steam injection is typically more environmentally sustainable than other techniques to enhance oil recovery, resulting in fewer issues during its use. The steam is not a significant concern, as it simply condenses into water, which does not produce pollution. Steam condenses in the well, transforming into water that then combines with oil. This results in increased operational expenses due to the need to manage bigger quantities of liquids. An additional dehydration facility is required nearby to ensure proper oil-water separation before oil transport. Steam injection may cause significant harm to the subsurface structure of the well. This occurs in reservoirs susceptible to geological rearrangements. This phenomenon has the potential to jeopardize the safety of workers and cause harm to the equipment. Thus, steam injection is not suitable for these reservoirs unless additional precautions are implemented. An economic difficulty is a key factor in the decision to utilize steam flooding. During the initial steam injection, one barrel of steam can recover additional oil of up to 30 barrels. Over time, the process becomes less efficient, resulting in 0.2 barrels of additional oil being recovered by one barrel of steam. At this stage, producers typically close the well until new technology is introduced. Figure 5.17 shows the schematic view of steam injection.

5.22.4 IN-SITU COMBUSTION (ISC)

This method for enhancing oil recovery is induced thermally, wherein the energy generation happens in situ by infusing an oxidizing gas, such as air or oxygen (O_2)-enriched air, to burn a portion of the heavy oil, which acts as a fuel, typically 5% to 10% of the OIP. This recovery method offers numerous benefits such as reducing steam-related expenses, significantly lowering greenhouse gas emissions, eliminating the necessity for water cycling procedures, upgrading heavy oil in place, and bypassing energy-intensive techniques later in the production process. These advantages render this method more ecologically sustainable and financially feasible. ISC techniques have garnered considerable attention for improving oil recovery. Three prominent techniques in this field include dry and wet forward combustion and reverse combustion. The methods entail the deliberate burning of oil in the subsurface formations to enhance oil flow and extraction. Figure 5.18 shows the schematic view of in-situ injection.

5.22.5 DRY FORWARD COMBUSTION (DFC)

This method includes injecting air into a specific well, then igniting oil either through auto-ignition or with the use of external sources of heat like electrical or gas heaters. After the oil ignites, various heat zones form in the reservoir as a result of heat and

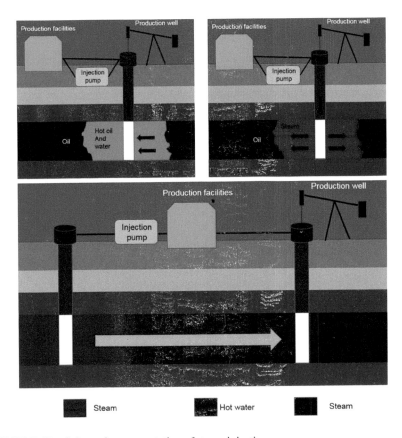

FIGURE 5.17 Schematic representation of steam injection.

mass transfer. These areas create unique temperature patterns. A combustion front forms in zones where some of the fuel (oil or coke) goes through the combustion process, producing heat. The heat is conveyed through convection in the water aiding in oil mobilization. A continuous air injection method is used to support the combustion front's movement in the direction of the production well, wherein the injected air and the combustion front travel in the same direction. In traditional DFC, air or O_2 is injected into the reservoir to ignite the coke, maintain the combustion front, and push the liquid hydrocarbons toward the producer well.

5.22.6 WET FORWARD COMBUSTION (WFC)

In DFC, only O_2 is supplied. However, most of the heat stays in the area behind the combustion front due to the low heat capacity of the gas. Alternatively, injecting air into water can enhance heat transmission. To address this problem, wet combustion (wet forward combustion, WFC) was developed to introduce heat to the area in front of the combustion front.

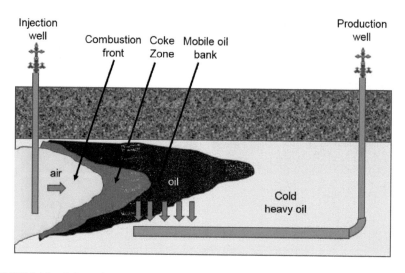

FIGURE 5.18 Schematic representation of in-situ injection.

5.22.7 Reverse Combustion

Reverse combustion (RC), also known as counter-current ISC, operates in a way where the combustion front starts close to the production well and advances towards the injector well. The oil is simultaneously pushed toward the producing well. This causes the combustion front and the air to move in opposite directions. This method was suggested for reservoirs of high-viscosity oils and tars in which hydrocarbons need to migrate from hot to cold areas, leading to decreased mobility and increased flow limitations. This approach enhances the flow of hydrocarbons during production while minimizing heat losses. ISC requires high temperatures to initiate and sustain the combustion process. Achieving and maintaining these temperatures while controlling and monitoring the combustion process in real time is challenging.

5.23 CHEMICAL FLOODING

Figure 5.19 shows a schematic view of chemical flooding.

5.23.1 Polymer Flooding

Polymer flooding is used to enhance oil recovery by improving the sweep efficiency of the reservoir capacity, which was previously inadequate during the water-flood field operations. Water-soluble polymers are injected to enhance sweep efficiency. The process is aimed at elevating the water's viscosity, reducing the occurrence of viscous fingering, enhancing the mobility ratio, and enhancing the overall efficiency on a large scale. In polymer flooding, fresh water is injected into the reservoir before injecting a slug of polymer solution in a quantity of 0.3 pore volume or higher. After

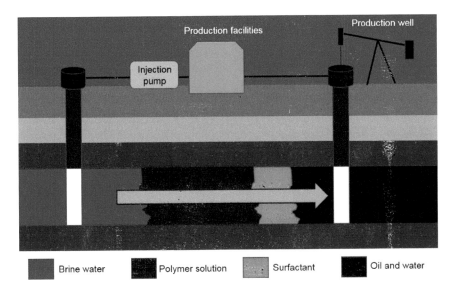

| | Brine water | | Polymer solution | | Surfactant | | Oil and water |

FIGURE 5.19   Schematic view of chemical flooding.

injecting the polymer slug, fresh-water slug is injected which is followed by continuous injection of drive-water.

The injection of polymer slug in between the injections of fresh water is intended
to reduce the contact with the saline water of the reservoir directly. The saline water
decreases the viscosity of the polymer solution. It enhances oil recovery compared to
water flooding by expanding the reservoir volume that is accessed, thereby raising the
macroscopic efficiency. Reservoirs with significant permeability fluctuations can be
remediated using polymer solutions. Reservoirs experiencing quick water-breakthrough
in the producing well may be subjected to polymer solution treatment. Polysaccharides
and polyacrylamides are commonly used polymers in this flooding process.

5.23.2   SURFACTANT FLOODING

In surfactant flooding, surfactants are injected in appropriate concentrations to
enhance microscopic efficiency and minimize the IFT. In this method, surfactants
are injected in appropriate concentrations based on the specific needs of the procedure. The method involves initiating the process with the fresh-water injection as a
pre-flush in surfactant flooding. Subsequently, surfactants are injected into the injector well. The polymer can be injected to enhance macroscopic efficiency. Ultimately,
the injection of drive water takes place. Surfactants for surfactant flooding are chosen
based on factors such as reservoir conditions, oil type, and desired outcome. This
method is incompatible with carbonate and high-salinity reservoirs and is too expensive. Common surfactants used in surfactant flooding include anionic surfactant

(sodium dodecyl sulfate or SDS), cationic surfactant (cetyl-trimethyl-ammonium bromide or CTAB), non-ionic surfactant (alkyl polyethylene glycol ethers, alkylphenol ethoxylates), and amphoteric surfactant (betaines and amine oxides).

Surfactant flooding is typically a component of a comprehensive enhanced oil recovery plan, which may involve injecting additional chemicals like polymers or alkalis to enhance displacement efficiency. The objective is to decrease residual oil saturation and improve the recovery of hydrocarbons from the reservoir. The success of surfactant flooding relies significantly on the specific conditions of the reservoir and the characteristics of the oil reservoir.

5.23.3 ALKALINE FLOODING

Alkaline flooding, commonly referred to as caustic flooding, is a procedure used to enhance the process of displacing oil. Because alkali, such as sodium hydroxide (NaOH) and potassium hydroxide (KOH) is used in this technique, it is more cost-effective compared to other EOR methods. Alkaline flooding involves two types of displacement mechanisms.

5.24 EMULSIFICATION

Emulsification occurs during the reaction of alkaline solutions with organic acids to create emulsifying soap, leading to an increase in the capillary number. Consequently, the IFT between oil and water decreases dramatically. The oil-water emulsion created is carried along by the flow of the fluid and may subsequently be extracted. This leads to a decrease in the amount of remaining oil and an increase in the amount of oil that can be extracted.

5.25 WETTABILITY REVERSAL

Displacement occurring in an oil-wet reservoir where oil is in the continuous phase hinders oil recovery. Alkaline compounds alter the injection water's pH, causing the rock wettability to shift from oil-wet to water-wet. This occurrence is known as wettability reversal. The relative permeability of the water-oil system and their mobility ratio drops, leading to an improvement in the oil recovery. The method involves injecting fresh water immediately as a pre-flush in alkaline flooding. Following this, an alkaline solution equivalent to 30% of the pore volume is injected into the injector well. Alkali is used, along with low-concentration surfactants and polymers to enhance sweep efficiency and regulate mobility. The mixture is then followed by the drive water. Alkaline flooding is not recommended for carbonate reservoirs due to the occurrence of calcium ions that react with alkalis to create calcium carbonate precipitates. These three components when combined are referred to as ASP flooding (alkaline-surfactant-polymer flooding)

5.26 GAS INJECTION

Gas injection is advantageous because it offers lower viscosity and higher injectivity compared to water. Injecting hydrocarbons or non-hydrocarbons into oil reservoirs is referred to as gas flooding. The injected gas may consist of hydrocarbons ranging from C_1 to C_5, or it may contain no hydrocarbons and instead include CO_2 and N_2. These gases exist as vapor at normal atmospheric pressure and temperature but can become supercritical fluids under reservoir conditions. For example, CO_2 exhibits a viscosity close to that of vapors but a similar density under reservoir conditions. CO_2 has a lower minimum miscibility pressure (MMP) compared to other gases like CH_4 and N_2. MMP stands for the minimum pressure required to recover 95% of the oil contacted. This parameter depends on both the composition of the crude oil and the temperature. The EOR technique primarily consists of N_2 flooding, CO_2 flooding, and gravity-assisted gas flooding. CO_2 flooding is commonly used to decrease carbon emissions and boost oil production.

There are two methods for recovering oil using gas injection processes.

- One of the primary aspects is the mass transfer process between the oil and the gas. Enhancing this mechanism involves increasing oil-gas miscibility.
- The second mechanism involves decreasing the oil viscosity and causing it to swell as the condensation of gas components happens in the oil. Oil recovery enhancement is achieved by maximizing the interaction between gas and the reservoir.

CO2 injection can occur in two ways:

In the first method, CO2 is injected into the reservoir to mix with the oil and decrease its viscosity in miscible CO2 flooding. This method is highly efficient in reservoirs where the injected CO2 and oil create a miscible solution, enhancing the oil displacement process. In the second method, CO2 is injected in liquid form into the reservoir, where it vaporizes and expands upon contact with the reservoir fluids. This expansion aids in displacing oil by generating pressure and decreasing viscosity.

The gases being injected are miscible, causing the trapped oil to mix with the injected gas. Hydrocarbons or injected gas are used to push the oil towards the production well. Fingering phenomena may arise in the gas flooding phenomena during the field applications. This could be caused by gravity override or reservoir heterogeneity. The injected gas flows through highly permeable areas, allowing some oil to be bypassed, resulting in low sweeping efficiency. Non-piston-like motion can also occur in homogeneous reservoirs. Both economics and physics play crucial roles in the design of the processes of gas flooding. Rock characteristics, fluid properties, phase behavior, and viscosity are important factors in the gas flooding process.

5.27 NEW EOR TECHNIQUES

Researchers and engineers are exploring and developing new EOR techniques to enhance the extraction of hydrocarbons from reservoirs as the oil and gas industry progresses. Some innovative and new EOR techniques include the following:

Nanoparticles in EOR: Nanoparticles like silica or polymer-based ones are being studied for their ability to change rock and fluid characteristics. They can be utilized to control wettability, modify fluid mobility, and improve sweep efficiency.

Microbial EOR: Microorganisms, like bacteria, can be added to the reservoir to improve oil recovery. Microbes can create biosurfactants and gases that change the reservoir's characteristics, enhancing oil flow.

Low-salinity water flooding: Modifying the salinity of injected water can impact the wettability of reservoir rocks, causing them to become more water-wet. This modification enhances water-rock interactions, decreasing residual oil saturation and improving recovery.

Smart-water injection: Altering the composition of injected water through the addition of particular ions or chemicals can change rock wettability and enhance oil recovery. This technique entails adjusting the water chemistry to suit the particular reservoir conditions.

Thermal EOR with electromagnetic heating: In addition to conventional steam injection and ISC, novel thermal EOR techniques are under investigation. Electromagnetic heating employs radio-frequency or microwave heating to lower oil viscosity.

EOR using ionic liquids: Researchers are studying ionic liquids, which are liquid salts at low temperatures, for their potential use as EOR agents. They can modify the IFT between oil and water, enhancing oil's mobility.

Elastic and smart polymers: Novel polymers, including elastic and smart polymers, are being created to enhance the injected fluid's viscosity. These polymers can undergo reversible alterations in their rheological characteristics, enabling improved management of fluid movement.

Foam EOR: Foam created by injecting gas and surfactants can help manage mobility and enhance sweep efficiency. Foam can hinder the movement of gas, redirecting it to the areas that have not been swept and enhancing the oil displacement.

The novel EOR techniques are currently in different phases of research, testing, and implementation. The efficacy of each approach is contingent upon the unique reservoir conditions, and successful execution typically necessitates a comprehensive understanding of the reservoir's geology, fluid properties, and rock characteristics.

5.28   OIL SPILL AND PREVENTION

The Deepwater Horizon (DWH) oil spill in 2010 questioned the existing beliefs of the industry, the research community, the government, and the public regarding marine oil spills (Murawski et al., 2020). It highlighted the significant technological challenges and risks involved in sustaining global hydrocarbon supplies. Presently, oil extracted from the Gulf of Mexico makes up over 90% of the United States marine oil output and around 20% of the nation's overall oil production from both land and sea. More than 50% of the marine-derived crude oil supply in the U.S. is currently sourced from wells above 1500 meters in ocean depth, categorized as ultra-deep production by the government and industry regulators. Technologies for extracting resources from ultra-deep and extremely productive formations have advanced quickly since 2000, a time when such wells were not present globally. Modern deep-water drilling utilizes ships with mooring systems having strings of drilling and production pipes running from the surface of the sea to the blow-out preventers (BOP) located on the sea floor, eliminating the need for derricks on the sea bottom. Ultra-deep exploration and production face challenges such as strong ocean currents, low temperatures and high pressures at sea-bottom, diverse sub-bottom rock and sediment layers, and high oil and gas reservoir temperatures and pressures (Pilisi et al., 2010). The combination of these characteristics, together with the exorbitant production expenses of ultra-deep wells, presents significant obstacles in properly exploring, developing, and operating facilities to extract hydrocarbons from ultra-deep settings while minimizing environmental harm.

Oil spills are the accidental discharge of crude oil or refined petroleum products into the environment, usually water bodies but can also occur on land. These spills can result in significant environmental, economic, and social repercussions. Oil spills can occur due to accidents in oil exploration, production, transportation (like tanker accidents), and refining. Natural disasters, including hurricanes or earthquakes, can cause oil leaks. Oil spills have the potential to damage both marine and terrestrial ecosystems. Oil can cover and harm plants and animals, interrupt food chains, and harm habitats. Avian species, marine mammals, fish, and other wildlife are especially susceptible to the harmful impacts of oil. Responding to oil spills includes containment, recovery, and clean-up measures. Booms, skimmers, and chemical dispersants are frequently employed to contain the oil and reduce its effects. Manual clean-up frequently requires the involvement of workers and specialized equipment. The repercussions of an oil spill can endure for an extended period following the initial occurrence. Despite a cleanup, certain regions may still suffer from ecological harm and disturbances in the ecosystem. Oil spills can cause substantial economic consequences for industries like fisheries and tourism. Coastal areas dependent on these businesses could encounter enduring economic difficulties as a result of the spill's impact. Marine oil spills affect beach tourism as well. For instance, the DWH oil spill affected tourism on the Gulf Coast, leading to the responsible parties having to provide compensation to those who suffered economic losses.

Oil is a term that encompasses a wide variety of hydrocarbon-based substances. Hydrocarbons are compounds consisting of hydrogen and carbon atoms. This encompasses items typically classified as oils, such as crude oil and refined petroleum products, along with other non-petroleum oils. Every kind of oil possesses unique physical and chemical characteristics. These characteristics impact the oil's spreading and degradation, its potential danger to human and aquatic life, and the probability of it posing a risk to natural and man-made assets. The rate of dispersion of an oil spill will have its environmental consequences. Many oils tend to disperse laterally, forming a sleek and smooth surface, known as a slick, on the water's surface. Viscosity, specific gravity, and surface tension are factors that influence the spread of an oil spill.

Viscosity is the degree of resistance of a liquid to flow. Increased oil viscosity results in enhanced immobility. Specific gravity is the ratio of the substance's density to water's density. Due to their lower density compared to water, most oils will naturally float on the surface of water. The oil spill's specific gravity may rise when lighter components in the oil evaporate. Viscous oils can sink and coalesce into tar balls or its interaction with the sediments or rocks at the water body's bottom may occur. Surface tension is the measurement of attraction between the molecules at the surface of a liquid. Greater oil surface tension increases the likelihood of a spill staying in one location and lower oil surface tension allows it to spread without assistance from wind or water currents. Higher temperatures can lower the surface tension of a liquid, causing oil to spread more easily in warmer water compared to very cold water.

Natural processes are constantly occurring in aquatic environments. These can mitigate the impact of the oil spill and expedite the restoration of the affected area. Natural processes include weathering, evaporation, oxidation, biodegradation, and emulsification.

- Weathering is a process involving physical and chemical alterations that result in spilled oil breaking down and becoming heavier than the surrounding water. Wave motion can cause natural dispersion by breaking up an oil slick into droplets that are then spread into the water column in the vertical direction. A thin layer or a secondary slick can be created on the water's surface by the previously mentioned droplets.
- Evaporation happens when the less dense or more volatile components in the oil blend turn into vapors and escape from the water's surface. This process separates the heavier oil components, which could experience additional weathering or settle on the ocean floor. Spills of lighter refined products like gasoline or kerosene consist of a significant number of combustible components, referred to as light ends. These substances may dissipate quickly, resulting in negligible impact on the aquatic ecosystem. Heavier oils provide a denser and more viscous residue. These oils have a lower likelihood of evaporating.
- Oxidation happens when oil comes into contact with water and oxygen reacts with oil hydrocarbons, creating water-soluble molecules. This mechanism mostly impacts oil slicks around their peripheries. Thick oils may only undergo partial oxidation, resulting in the formation of tar balls. These compact, adhesive, black spheres can persist in the environment, appearing on coastlines long beyond the spill.
- Biodegradation is the process in which microorganisms, including bacteria, consume oil hydrocarbons. A diverse array of microorganisms is necessary to achieve a substantial decrease in the oil content. To support biodegradation, nutrients like phosphorous and nitrogen are occasionally introduced into the water to stimulate the growth and reproduction of microorganisms. Biodegradation is most effective in warm-water settings.
- Emulsification is the process of creating emulsions, which consist of minute droplets of oil and water mixed. Emulsions are created from wave action and significantly impede weathering and cleanup procedures. Water-in-oil emulsions, often referred to as chocolate mousse, are created when intense wave

action results in water being enclosed within thick oil. Chocolate mousse emulsions can persist for an extended period, ranging from months to years, in the environment. Oil-in-water emulsions make oil sink and vanish from the surface, creating the illusion that it has disappeared and the environmental danger has been eliminated.

Natural processes vary between freshwater and saltwater habitats. Environmental consequences in freshwater settings can be of higher severity due to reduced water flow. Oil accumulates and persists in stagnant water bodies for extended durations. Oil accumulates on vegetation along the banks of streams and rivers. Oil can also impact the sediments existing in the bottom of the freshwater bodies, influencing creatures that inhabit or rely on sediments. Controlling oil spills involves two main steps: containment and recovery. It is crucial to promptly control an oil spill on water to reduce risks and prevent harm to individuals, assets, and the environment. Containment equipment is utilized to contain the oil spill and facilitate its removal, recovery, or dispersion.

Floating barriers, known as booms, are the primary equipment used to contain the spread of oil. Containment booms are utilized to manage the spread of oil to minimize the risk of contaminating shorelines and other resources. They also help to gather oil in denser surface layers, facilitating the recovery process. Booms can be utilized to redirect and guide oil slicks down certain routes, facilitating their removal from the water's surface. Despite variations in form and construction, booms typically have four fundamental features.

- Floatation device.
- Above-water free-board: Used to control the oil and to avoid the splashing of oil by the waves onto the boom.
- Underwater skirt: Designed to confine oil and minimize oil spillage beneath the containment boom.
- Longitudinal support: Used to reinforce the boom against the forces of wind and wave and to enhance the stability and maintain the boom in an upright position.

Table 5.1 summarizes the types of booms and their vital characteristics.

Booms often function effectively in calm seas with regular, elongated waves; however, turbulent and choppy water conditions are prone to causing boom malfunctions. Wave, wind, and current loads can weaken a boom's capacity to contain oil. Entrainment is the loss of oil that happens when friction between water and oil causes oil droplets to detach from the slick and are sucked under the boom. If a leak happens and there is no containment equipment, makeshift barriers can be created using available materials. Improvised booms are commonly employed as temporary solutions to contain or redirect oil until advanced technology is available. Figure 5.20 shows view of boomers.

After an oil spill is contained, the process of removing the oil from the water may commence. Booms, skimmers, and sorbents are the three types of equipment used in this process. Booms used in oil recovery are typically supported by a horizontal arm

TABLE 5.1

Types of Booms and their Characteristics

Type of Booms	Characteristics
Fence	High free-board
	Flat floatation device
	Less useful in turbulent water conditions
	Subjected to boom twisting due to wave and wind forces
Round or curtain	Circular floatation device
	Continuous skirt
	Excel in turbulent water conditions
	Require more effort to clean and maintain
Non-rigid inflatable	Available in many forms
	Excel in rough sea
	Easy to clean and maintain
	Costly, intricate to operate, and prone to punctures and deflation

that extends from one or both sides of a vessel. The vessel moves slowly through the densest parts of the spill, collecting and containing the oil between the boom's angle and the vessel's hull. In another scenario, a boom is anchored at the extremities of a stiff arm protruding from the ship, creating a J- or U-shaped enclosure where oil can accumulate. The trapped oil can be extracted and sent to storage tanks for disposal or recycling.

A skimmer is a tool used to retrieve the spilled oil. Skimmers' efficiency is contingent upon weather conditions. They recover more water than oil in moderately turbulent or choppy water. The three types of skimmers are weir, oleophilic, and suction.

- Weir skimmers utilize a barrier or structure placed at the boundary between oil and water. The oil that floats on the water will pour over the dam and be contained in a well, minimizing the amount of water it carries. The oil and water mixture can be extracted using a pipe or hose and transferred to another storage facility for being recycled or disposed of. These skimmers are susceptible to getting stuck and obstructed by floating material.
- Oleophilic skimmers utilize disks, belts, or mop chains made of oleophilic materials to absorb oil from the water surface. The oil is extracted by squeezing or scraping it into a recovery tank. Oleophilic skimmers are advantageous due to their versatility, enabling them to properly handle any spills irrespective of their thickness. Chain-mop skimmers are effective on water that faces obstruction from rough ice or debris.
- A suction skimmer functions similarly to a domestic vacuum cleaner. Oil is extracted through large floating heads and transferred to storage tanks using pumps. Suction skimmers are highly efficient yet prone to clogging by debris and need continuous skilled monitoring. Suction skimmers perform most well in calm waters where oil has accumulated against the barrier. Figure 5.21 shows skimmer.

FIGURE 5.20   Boomer.

Sorbents absorb liquid and can be utilized for oil recovery using adsorption, absorption, or a combination of both. In the case of adsorbents, the oil is attracted to the substance's surface, while in the case of absorbents, the oil is absorbed into their pore spaces (Cano and Dorn, 1996). For sorbents to effectively counteract oil spills, they must possess both hydrophobic and oleophilic properties. Sorbents are primarily utilized to eliminate remaining oil residues or in inaccessible regions, although they can also be employed as the primary cleanup procedure for minor spills. Appropriate disposal or recycling must be done to the oil extracted from the sorbent materials. Sorbent materials are of three types.

- Natural organic: They are affordable and easy to find. They can absorb oil ranging from three to 15 times their weight, although they have certain drawbacks. Certain substances tend to absorb both water and oil, leading to their sinking. Several organic sorbents are porous. It is quite challenging to gather them once they are dispersed over the water. For example, sawdust, hay, straw, peat moss, corncobs, etc.
- Natural inorganic: They can absorb oil ranging from four to 20 times their weight. They are cost-effective and abundantly accessible. For example, volcanic ash, sand, wool, glass, vermiculite, perlite, clay, etc.
- Synthetic: They have the capacity to absorb oil up to 70 times their weight and certain varieties may be cleaned and reused multiple times. Non-reusable synthetic sorbents offer challenges as they require temporary storage before proper disposal. For example, artificial materials resembling plastics, like nylon fibers, polyethylene, polyurethane, etc.

While selecting sorbents for spill cleanup, one must consider the following characteristics.

FIGURE 5.21   Skimmer.

FIGURE 5.22   Sorbents.

- The absorption rate fluctuates depending on the oil's thickness. Light oils are absorbed faster than heavier oils.
- Oil retention can lead to sagging and deformation of a sorbent structure due to the weight of the recovered oil. When removed from water, the materials expel the oil that is confined in its pores. Lighter, less viscous oil is lost more easily via the pores than heavier, more viscous oil during the recovery of absorbent materials.
- Sorbents can be applied to spills either manually or mechanically, utilizing blowers or fans. Several natural inorganic sorbents, such as vermiculite and clay, are powdery, challenging to use in windy environments, and may pose a risk if breathed. Figure 5.22 shows the sorbent.

6 Subsea Systems

Summary

This chapter gives a brief summary of subsea systems, explaining the glossary of components. Detailed illustrations on subsea system layout and subsea manifold help the readers to understand the location importance and nature of their function in the subsea system. Production tree with the key role of jumpers and umbilicals are also discussed in detail.

6.1 SUBSEA PRODUCTION SYSTEMS (SSS)

Subsea systems (SSS) are multicomponent seabed systems that enable the production of hydrocarbons in water depths where conventional fixed or bottom-founded platforms cannot be installed (Bai, 2001). Subsea systems can be located several kilometers away in deeper water and tied back to existing host facilities like FPS. The constituent parts of SSS are (a) array of subsea wells, (b) manifolds, (c) central umbilical, and (d) flow lines. The different layout arrangements of subsea systems configurations are single-well satellite, multi-well satellite, cluster well system, template system, and combinations of those previously mentioned. The various components are organized as follows:

6.1.1 Subsea Production Tree

It is an arrangement of valves, pipes, fittings, and connectors placed on top of a wellbore. They are typical for all types of drilling platforms. The height varies about 30 m, requiring a counterweight to regulate the flow under pressure. The orientation of the valve can be done in the horizontal bore or the vertical outlet of the tree; comparatively, the horizontal outlet is safe. The valves can be operated by hydraulic signals or electrical or manually by the diver to activate them.

6.1.2 Pipeline and Flow Line

These are the conduits to transport fluid from one place to another. Pipelines are utilized to move a large region for transporting oil, gas, sulfur, and produced water from two separate facilities. The length range varied from 1 m to 1000 km, and the pipeline's diameter is about 450 mm. The flow lines are installed within the confines of the platforms or manifold boundaries. These diversion lines are used to route the subsea manifold into the processing machinery (Guo et al., 2005).

6.1.3 Umbilical

It connects the host and the subsea system. These are the bundled arrangement of tubing, piping, and electrical conductors of 25 mm diameter are packaged together.

DOI: 10.1201/9781003497660-6

An armored sheath is provided for insulation. It transfers the control fluid and electric current to regulate the operations of the subsea production and safety equipment. These are the specifically designed dedicated tubes to monitor pressure and inject fluids.

6.1.4   JUMPER

These are the check arrangements for the umbilical used to connect several subsea equipment. The offset distance between the components governs the jumper's length. They are categorized into the following three categories: (a) production jumper, (b) hydraulic jumper, and (c) electrical jumper. They are provided at the terminal. A flexible jumper system provides versatility, unlike a rigid jumper system that restricts the space and the handling capacity.

6.1.5   TERMINATION UNIT

It is equipment designed to facilitate the interaction between the subsea manifold and the pipeline or flow line more quickly. It is positioned near the manifolds and can be used for electric or hydraulic control. It has an installation arm to brace it during the lowering process.

6.1.6   PRODUCTION RISERS

The flow line component resides between the host facility and the seafloor adjacent to the host. The length of the riser depends upon the water depth, and the riser configuration is as follows: (i) free hanging, (ii) lazy S, (iii) lazy wave, (iv) steep S, and (v) steep wave. They are designed to resist all types of forces.

6.1.7   TEMPLATE

The subsea equipment is housed in a fabricated structure. It accommodates multiple trees in tight clusters, manifolds, piping equipment, and chemical treatment equipment.

Figure 6.1 shows different layout of subsea systems: a single-well satellite, a multiwell satellite, a cluster-well satellite, a template, and a combination of those previously mentioned. A multi-component system comprises a subsea production tree, pipeline and flow line, subsea manifold, umbilical, host facility, termination units, production risers, templates, and jumpers.

Subsea systems are multicomponent seabed systems. They allow the production of hydrocarbons in water depths where conventional fixed or bottom-founded platforms cannot be installed. They comprise an array of subsea wells, a variety of manifolds, and a central umbilical. A subsea production tree is an arrangement of valves, pipes, fittings, and connections placed on top of a wellbore. Orientation of the valves can be in the vertical bore or the horizontal outlet of the tree valves, which electrical or hydraulic signals can operate; alternatively, they are also operated by a remotely operated vehicle (ROV). A schematic view of the valve, operated by the ROV, is shown in Figure 6.2.

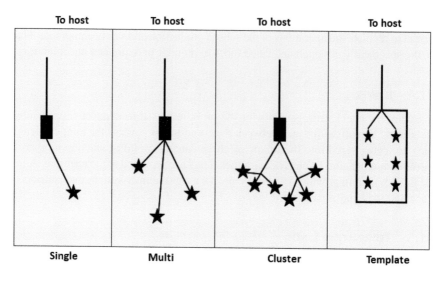

FIGURE 6.1   Different layouts of subsea systems.

FIGURE 6.2   Valve operated by an ROV.

6.2 SUBSEA MANIFOLD

These are gravity-based, stand-alone, or integrated structures containing valves, pipelines, and fittings. It serves as a central gathering point for production from subsea wells. The combined flow is redirected to the host facility. They are rectangular or circular with a height of about 9 m. These are anchored to the seafloor by piles. The size of the manifold depends upon the pattern layout and number of wells. Figure 6.3 shows a typical subsea manifold. It is a gravity-based seafloor structure that consists of valves, pipelines, and fittings. It serves as a central gathering point for production from subsea wells and redirects the combined flow to the host facility. For example, a subsea manifold may not be needed for all subsea designs in field developments where individual production trees are directly tied to the host facility. A manifold arrangement can be any shape but usually is circular or rectangular. Generally, it is either a stand-alone structure or integrated into a well template. The manifold may be anchored to the seafloor with piles or skirts penetrating the mud line. Although the number of wells governs the size of the manifold, its pattern depends on how the wells are integrated into the system. A typical subsea manifold will have dimensions of 24 m in diameter and 9 m high above the seabed.

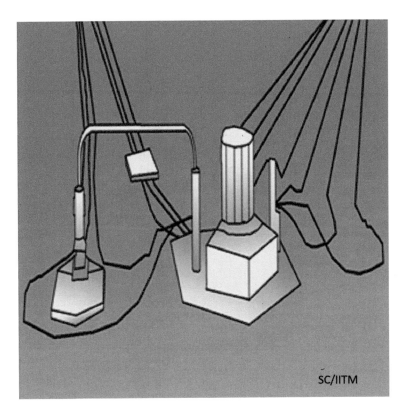

FIGURE 6.3   Subsea manifold.

6.3 PRODUCTION TREE

The arrangement of the valves in the production tree depends on the type of tree: vertical bore or horizontal bore. Figure 6.4 shows the type of production trees. Pipelines and flow lines, part of the multicomponent system, are conduits for transporting the fluid from one region to another. Pipelines are piping, risers, and appurtenances set installed for the purpose of transporting oil, gas, sulfur, and produced waters between two separate facilities. The length and size of a pipeline or flow line depend on its purpose. Pipe lengths can range from 1 m to 100 km and are typically 450 mm in diameter. Flow lines are piping installed within the confines of the platform or manifold. They are installed for the purpose of mixing the subsea manifold or routing into the processing equipment.

6.4 JUMPERS AND UMBILICALS

A schematic view of the jumpers and umbilical is shown in Figure 6.5. An umbilical is a bundled arrangement of tubing, piping, and/or electrical conductors in an

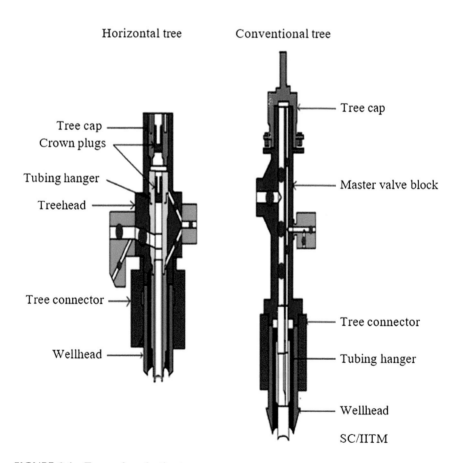

FIGURE 6.4   Types of production tree.

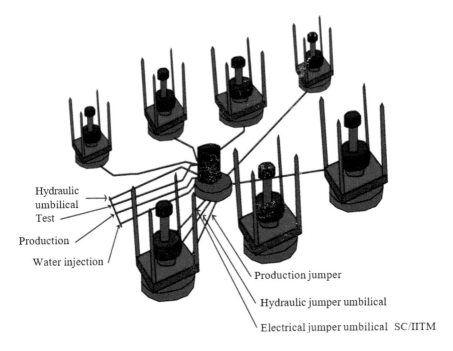

Hydraulic
umbilical
Test
Production
Water injection
Production jumper
Hydraulic jumper umbilical
Electrical jumper umbilical SC/IITM

FIGURE 6.5 Umbilicals and jumpers.

armored sheath, which is installed from the host facility to the subsea production
system. An umbilical is used to transmit the control fluid and/or electrical current
necessary to control the functions of the subsea production and safety equipment
(tree, valves, manifolds, etc.) (Nordic Committee for Building Regulations, 1977;
Norwegian Petroleum Directorate, 1985; OCS, 1980). Dedicated tubes in an umbilical are used to monitor pressures and inject fluids (chemicals such as methanol) from
the host facility to critical areas within the subsea production equipment.

Electrical conductors transmit the power to operate subsea electronic devices. The
diameter of the dimensions typically ranges up to 200 mm. The umbilical includes
multiple tubes, typically ranging in size up to 25 mm. The number of tubes depends
on the complexity of the production system. The length of an umbilical is defined by
the spacing of the subsea components and the distance these components are located
from the host facility. A typical host facility can be any of the various platforms used
for developing offshore hydrocarbon resource fields, including fixed jacket-type platforms, TLPs, spars, FPSs, FPSOs, and off-loading systems. The type of host facility
used for the subsea production system depends on the water depth, the type of field
development, the reserve base, and the distance from the infrastructure, but it is also
largely governed by economic factors. Figure 6.6 shows jumpers.

6.5 TERMINATION UNIT

The termination unit facilitates the interface of the umbilical, pipeline, or flow
line with the subsea equipment. It has several analogous names, including pipeline

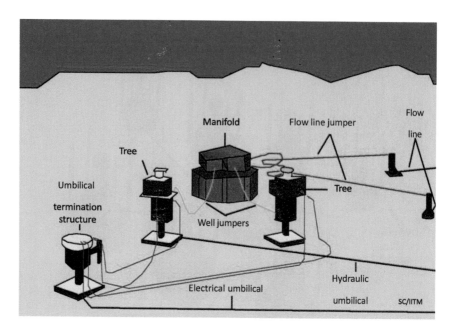

FIGURE 6.6 Jumpers.

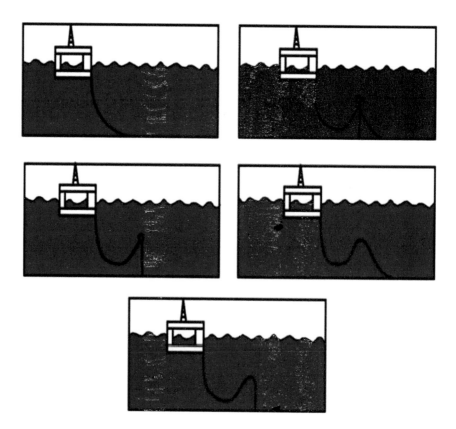

FIGURE 6.7 Layouts of risers.

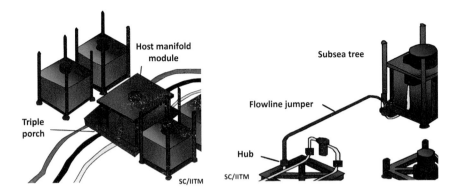

FIGURE 6.8 Template on subsea equipment.

end manifold, umbilical termination assembly, electrical distribution structure, and flow line lay-down sled. It can be used for electric and/or hydraulic control applications. It is generally equipped with an installation arm to brace it during the lowering process. It is positioned near subsea manifolds, production trees, and flow lines, alternatively incorporated into the design of manifolds and templates (Llyod's Register, 2005).

6.6 PRODUCTION RISER

A production riser is a portion of the flow line that resides between the host facility and the seabed adjacent to the host facility. They are usually 3–12 inches in diameter, whose length is governed by the water depth and riser arrangement configuration. Risers can either be vertical or assume a variety of waveforms; they can be either flexible or rigid. They can also be contained within the area of a fixed platform or floating facility, run on the seabed, as well as run partially in the water column. Figure 6.7 shows the different layouts of risers. A template is a fabricated structure that houses the subsea equipment. Templates can be of any shape but are typically rectangular. Dimensions range from 10 to 150 in. length, 10 to 70 in. width, and about 10 to 70 in. height. Templates can accommodate multiple trees in tight clusters, manifolds, pigging equipment, termination units, and chemical treatment equipment. Figure 6.7 shows a schematic layout of a template on subsea equipment. Jumpers are pipe spools typically ranging up to 0.5 m in diameter and 45 m in length. They are used to connect various subsea components. They are beneficial when connected to satellite wells through connections of small diameter production lines (3–6 inches), well testing lines (3–6 inches), hydraulic fluid lines (1 inch), and chemical service lines (1 inch) to the manifold. The offset distance between the components (trees, flow lines, manifolds, etc.) governs the jumper length and characteristics. Flexible jumper systems provide versatility, unlike rigid jumper systems, which limit the space and handling capability. Figure 6.8 shows a template of subsea equipment.

Exercise

1. What do you understand by the term offshore structures?
   Structures that do not have direct access from the coast are termed offshore structures. They are primarily deployed for oil and gas exploration.

2. When does a dynamic analysis become important?
   When structures are subjected to loads that vary with time, one needs to perform dynamic analysis. Further, dynamic analysis can be carried out only for mass-dominant structural systems.

3. List a few environmental loads that act on ocean structures.
   Wave load, current load, wind load, ice load, and impact loads from ships and supply vessels are a few of the crucial loads that act on ocean structures.

4. What is the specialty of the design of ocean structures?
   The design is different from the conventional ones as the members do not derive resistance from the strength of the materials but from their geometry. They are called form-dominant structures. One similar example of land-based structures, which are form-dominant, is trusses.

5. Name any one design code that governs the analysis, design, and construction of ocean structures.
   American Petroleum Institute, API. There are many API codes, classified as per the type of offshore structure.

6. Offshore structures are unique in design and _____, compared with other kind of structures (geometric form).

7. TLP is termed as a _____structure because it has two groups of natural frequencies; namely, flexible and rigid (hybrid).

8. What are the failure modes of gravity base structures?
   Sliding, bearing capacity failure, rocking, liquefaction.

9. Why are jacket structures also called template structures?
   The jacket acts as a guide for pile-driving.

10. Why are steel skirts used at the bottom of GBS?
    Avoids sliding and provides stability to the structure.

11. Differentiate riser and conductor.
    Risers are long, slender tubes that carry crude oil or partially processed oil to another location for further processing.

    Conductors are long, hollow, straight, or curved tubes that embed into the seabed through which drilling is performed.

12. Guy lines restrains .... and .... motions of tower (surge, sway).

13. The spud can offer a support connection, which is position-fixed and .... (rotation-free).

14. Articulated towers are offshore platforms that are connected to the seabed using .... (universal joints).

15. Explain the structural action of TLP with a neat sketch.

A tension leg platform (TLP) is designed with excess buoyancy, in comparison with its weight. For the basic form, whose weight is much lower than the buoyancy force, the platform will have a tendency to be pushed up when it is installed. Exceedance of the buoyancy force will be counteracted by imposing pretension in tethers. These tethers are used to anchor the platform to the seabed. It consists of large-diameter pontoons and column members, which are helpful in excessive buoyancy as the displaced volume will exceed its weight (since the pontoons and columns are hollow tubes). Because the legs of the platform will be imposed with high pretension, the name tension leg is associated with the platform. Commissioning of the platform is more straightforward, in comparison with the earlier structural forms of offshore structures. Because the buoyancy exceeds the weight of the platform, it remains free-floating.

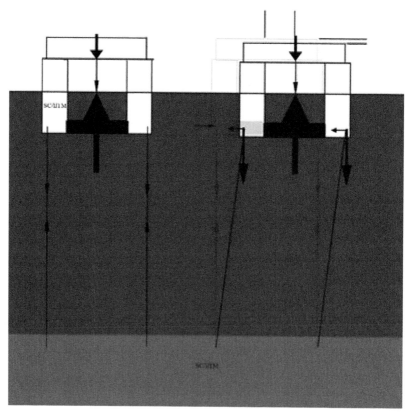

16. Plot the following: PM spectrum, modified PM spectrum, ISSC spectrum, and Jonswap spectrum. Compare the spectral plots, in terms of energy content and peak frequency.

Wave spectrums:

- Pierson-Moskowitz (PM) spectrum

It is a one-parameter spectrum used for fully developed sea conditions. It includes only the peak frequency and is given by:

$$S^+(\omega) = \frac{\alpha g^2}{\omega^5} \exp\left[-1.25\left(\frac{\omega}{\omega_0}\right)^{-4}\right]$$

- ISSC spectrum (International Ship Structures Congress)

$$S^+(\omega) = 0.1107 H_s \frac{\omega^{-4}}{\omega^5} \exp\left[-0.4427\left(\frac{\omega}{\bar{\omega}}\right)^{-4}\right]$$

$$\bar{\omega} = \frac{M_1}{M_0}$$

This spectrum is a modified form of the Bretschneider spectrum used for fully developed sea conditions. The governing equation of the sea state is given by:

- JONSWAP spectrum (Joint North Sea Wave Project)

It is a modified form of the PM spectrum recommended for reliability analysis. It is used for describing the winter storm waves in the North Sea. The governing equation of the sea state is given by:

$$S^+(\omega) = \frac{\bar{\alpha} g^2}{\omega^5} \exp\left[-1.25\left(\frac{\omega}{\omega_0}\right)^{-4}\right] \gamma^{a(\omega)}$$

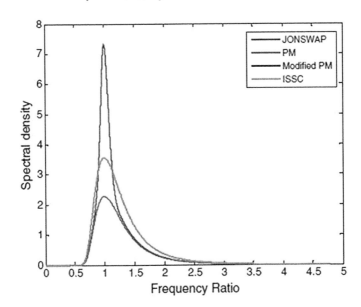

17. Explain DPS and its application in the offshore platform.

The DPS controls platform displacements in all the horizontal degrees of freedom. It is composed of a controller, a sensor system, a thruster system, and a power system. The sensor system feeds the controller with information about the platform positioning and environmental parameters that arise from wind, current, and waves. The controller commands the action of thrusters installed on the bottom of the platform hull, which, in turn, generates forces and moments needed to counteract the environmental forces. It helps to keep the vessel at the reference point. It keeps the platform within a tolerance radius of about 2%–6% of the water depth.

18. How will you estimate wind force on an offshore deck?

19. Discuss briefly different analytical concepts available for estimating wave forces on offshore structures.

20. Draw a neat sketch to explain different wave parameters considered to estimate wave forces.

21. Write a brief note on Airy's wave theory.

22. Write a brief note on Stokes' fifth-order wave theory.

23. How are waves classified according to relative water depth?

24. Explain, briefly, the wave-structure interaction.

25. How can we estimate the maximum wave force on an offshore structure?

26. What do you understand by buoyant force? How do you estimate it?

27. Why are current forces significant in offshore structures?

28. Write a brief note on i) earthquake forces; ii) snow/ice loads; and iii) accidental loads.

29. What do you understand by marine growth? What is its significance in the design of offshore structures?

30 How the effect of current is accounted for in wave loads? (Hint: Dopplers' effect.)

Problem 1:

A pile of diameter, $D = 0.75$m is to be installed in a water depth of 100 m. Wave height H and period T are 6m and 10 sec, respectively. Take $C_D = 1$, $C_M = 2$, and the density of seawater is 1020 kg/m^3. Compute the maximum wave force and moment at the base of the pile.

Data given:
pile diameter $D = 0.75$m
water depth $h = 100$m
wave height $H = 6$m
wave period $T = 10$s

Solution:

Deep water wavelength is given as $L_0 = 1.56 \times T^2 = 156 m$

$$\frac{h}{L_0} = \frac{100}{156} = 0.64 > 0.5 \text{ Hence, it is a deep-water condition.}$$

Hence, wavelength L is equal to deep water wavelength 156 m.

Wave number $k = \dfrac{2\pi}{L} = 0.04$

The phase angle at which the total wave force is maximum is given as

$$\theta_{max} = \sin^{-1}\left[\pm \frac{\pi D}{H}\frac{C_M}{C_D}\frac{2\sinh^2 kh}{2kh+\sinh 2kh}\right]$$

$$\theta_{max} = 51.3°$$

The maximum force is given by:

$$F = \rho g V \left(\frac{H}{2h}\right)\tanh kh \times \left[C_M \sin\theta + C_D\left(\frac{H}{4\pi D}\right)\frac{2kh+\sinh 2kh}{\sinh^2 kh}|\cos\theta|\cos\theta\right]$$

$$F_{max} = 27311.5 \text{ N}$$

where $V = \dfrac{\pi}{4}D^2 h$. Maximum overturning moment at the base of the pile is given by:

$$M = \rho g V \left(\frac{H}{2h}\right)\left(\frac{1}{k}\right)\tanh kh\left[C_M\frac{kh\sinh kh - \cosh kh + 1}{\sinh kh}\sin\theta + C_D\left(\frac{H}{2\pi D}\right)\right.$$
$$\left.\frac{(kh)^2 + kh\sinh 2kh - \sinh^2 kh}{\sinh^2 kh}|\cos\theta|\cos\theta\right]$$

$$M = 2.7 \times 10^6 \text{ Nm}$$

The distance at which the maximum force acts from the base of the pile

$$Y = \frac{2.7 \times 10^6}{27311.5} = 98.86 m$$

Problem 2:

Consider the front view of the offshore structure as shown in Figure 7.3 and determine the forces exerted on member 1-2 for relatively uniform wave-induced water motion over described by u = 4 m/s, v = 1.2 m/s, $a_x = 1.2$ m/s² $a_y = -1.6$ m/s² $C_D = 1$, $C_M = 2$ and density is 1025 kg/m³. The still water level above the seafloor is 29m. The diameter of the member is 0.6 m.

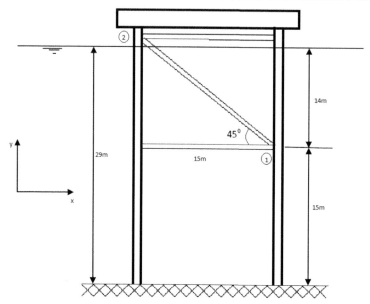

Solution:
The orientation of members is given as

$$\theta = 90° \quad \emptyset = 135°$$

$$C_x = \sin\emptyset \, \cos\theta = 0$$

$$C_y = \cos\emptyset = -0.707$$

$$C_z = \sin\emptyset \sin\theta = 0.707$$

$$V = \left[u^2 + v^2 - \left(C_x u + C_y v \right)^2 \right]^{1/2}$$

$$u_n = u - C_x \left(C_x u + C_y v \right) = 4m/s$$

$$v_n = v - C_y \left(C_x u + C_y v \right) = 0.6m/s$$

$$w_n = -C_z \left(C_x u + C_v \right) = 0.6m/s$$

$$a_{nx} = a_x - C_x \left(C_x a_x + C_y a_y \right) = 1.2m/s^2$$

$$a_{ny} = a_y \quad C_y \left(C_x a_x \mid C_y a_y \right) = \quad 0.8 \ m/s^2$$

$$a_{nz} = -C_z \left(C_x a_x + C_y a_y \right)$$

The magnitude of the velocity normal to the member is given by:

$$V = \left[4^2 + 1.2^2 - 0.707 \times 1.2 \right]^2 1/2 = 4.08 \ \text{m/s}^2$$

$$F_x = \frac{1}{2}\rho C_D D V u_n + \rho C_m \frac{\pi}{4} D^2 a_{nx}$$

= 0.5 × 1025 × 1 × 0.6 × 4.08 × 4 + 1025 × 2 × 3.14/4 × 0.62 × 1.2
= 5713.9 N/m

$$F_y = \frac{1}{2}\rho C_D D V v_n + \rho C_m \frac{\pi}{4} D^2 a_{ny}$$

= 0.5 × 1025 × 1 × 0.6 × 4.08 × -0.6 + 1025 × 2 × 3.14/4 × 0.62 ×-0.8
= 289 N/m
$F_z = F_y = 289$ N/m

where F_x, F_y, and F_z are force per unit length. The total force is calculated by multiplying the above force by the length of the member which is $29 \times \sqrt{2}$; this is valid as the following is assumed to be uniform.

Problem 3:

Figure 7.4 shows a simple 2D offshore structure with two piles and one diagonal member. Find the total force on the structure. The wave height and wavelength are given as 6 m and 90 m in a water depth of 25 m. The pile has a diameter of 1.2 m and the diagonal member has a diameter of 0.6 m. Assume that $C_D = 1$ and $C_M = 2$.

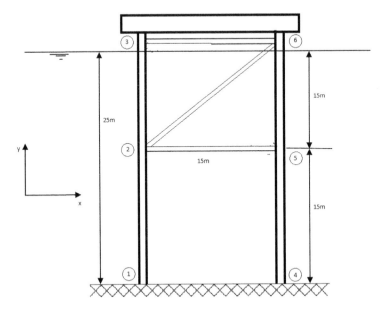

Solution:

Data given:

Wave height H = 6m

Wavelength L = 90m

Wave number $K = \dfrac{2\pi}{L} = 0.06$

Wave frequency $\omega = \sqrt{gk\ \tanh kd} = 0.73$ rad/s

Water depth h = 25m

For members 1-3
The total drag force is given by:

$$F_{D1-3} = \frac{\rho C_D D}{32k}(\omega H)^2 \left(\frac{\sinh 2ky}{\sinh^2 kh} + \frac{2ky}{\sinh^2 kh} \right)|\cos\theta|\cos\theta$$

where

$y = h + \eta = 25 + 3\cos(kx - \omega t)$

$\theta = kx - \omega t$

From subsequent trials, one can find that the wave force is maximum at $\omega t = 6$ and the maximum drag force is given by:

$$F_{D1-3} = 44394.4N$$

Total inertia force is given by:

$$F_{I1-3} = \frac{\rho C_M}{2k}\frac{\pi}{4}D^2\omega^2 H\frac{\sinh ky}{\sinh kd}\sin(kx - \omega t)$$

$$F_{I1-3} = 22043N$$

Hence, the total wave force on the pile 1-3 is given by:

$F_{T1-3} = F_{D1-3} + F_{I1-3} = 44394.4 + 22043 = 66437.4N$

For members 4-6, x = 15m
The total maximum inertia and drag forces are calculated as

$$F_{D4-6} = 6677N$$

$$F_{I4-6} = 78966N$$

$$F_{T4-6} = 85643N$$

Forces on members 2–6

For members 2–6, $\theta = 0$ and $\emptyset = 45^0$

$C_x = \sin\emptyset\cos\theta = 0.707\ C_x = \cos\emptyset = 0.707\ C_z = \sin\emptyset\sin\theta = 0$

The horizontal force per unit length acting on a side face diagonal is given by:

$$f_x = \frac{1}{2}\rho C_D D V u_n + \rho C_M \frac{\pi}{4}D^2 a_{nx}$$

where

$$u_n = u - C_x\left(C_x u + C_y v\right)$$

$$= \frac{1}{2}(u - v)$$

Similarly,

$$a_{nx} = \frac{1}{2}\left(a_x - a_y\right)$$

$$V = \sqrt{\left[u^{2+}v^2 - \left(C_x u + C_y v\right)^2\right]}$$

For members 2–6, we have both x and y varying along the member to its intersection with the water surface

$$y = 25 + 3\cos\left(kx - \omega t\right)$$

$$x = 10 + \cos\left(kx - \omega t\right)$$

For $\omega t = 6$ by trial, we get $x = 13m$

Hence, the elevation of the water surface on the member is 15 + 13 = 28 m and the length of the member struck by the wave is 13/0.707 = 18.038 m. Wave force per unit length varies along the member and we estimate the total force by dividing the member length struck by the wave into two segments each of 9.19 m. Wave force f_x is calculated at the mid-length of each section and the total force is computed by assuming these values, to be constant over the respective segments.

x(m)	y(m)	u(m/s)	v(m/s)	a_x(m/s²)	a_y(m/s²)	f_x(N/m)
3.24	18.24	1.51	0.62	0.573	-0.88	508.6
3.27	24.27	1.5535	1.5525	1.2636	-1.017	661

$$F_x = (508.6 + 661)9.19 = 10748.62N$$

The total maximum force on the complete structure is given by:

$$F_T = 66437.4 + 85643 + 44394.4 = 196472.8N$$

Problem 4:

Find the total horizontal force on the member 1–2 as shown in the figure and the moment at the base of the member due to this total horizontal force. The member is inclined at an angle of 30^0 to the vertical. Wave height and wave period are 6 m and 10 sec. The member diameter is 1.2 m, and the depth of

water is 100 m. Assume $C_D = 1, C_M = 2$, and density of seawater as 1025 kg/m³.
Also, plot the variation of total force with time.

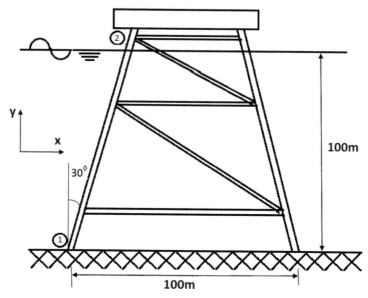

Solution:

Given:
pile diameter D=1.2m
water depth h = 100m
wave height H=6m
wave period T=10s
Deep water wavelength is given as $L_0 = 1.56 \times T^2 = 156$ m

$$\frac{h}{L_0} = \frac{100}{156} = 0.64 > 0.5 \text{ Hence, it is a deep-water condition.}$$

Wavelength L is 156m for deep water condition.

Wave number $k = \dfrac{2\pi}{L} = 0.04$

The total force on a pile for a segment dy is given by:

$$dF = \frac{1}{2}\rho C_D D |V_n| V_n dy + C_M \rho \frac{\pi D^2}{4} a_n dy$$

where
$V_n = u\cos\alpha + v\sin\alpha$

$a_n = a_x\cos\alpha + a_y\sin\alpha$

the $\alpha =$ inclination of the pile to vertical

uhorizontalhorizontal horizontal and vertical particle velocity from airy wave
theory

a_x, a_y = horizontal and vertical particle acceleration from airy wave.

The total force F on the member is calculated by integrating the segment force dF along the length of the pile. The variation of the total force F with time is shown as follows. Corresponding Matlab coding used for computation is also given at the end.

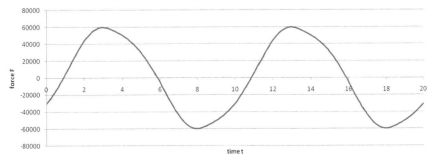

From the graph, it has been found that the total force is maximum at t = 3.3 sec and the corresponding value total force is 59771.77N. The moment at the base due to the force on segment dy is obtained as

$$dM = dFy$$

The moment due to total force is calculated by integrating the segment force moment along the length of the member. The moment at the base of the member due to this maximum total force is 4919.544kN-m. The distance at which this maximum total force acts is 82.3m (= 4919544/59771.77) = 82.3 m from the bottom.

Matlab code

```
clc
close all
clear all
H=6;
h=100;
h1=100/cosd(30)
D=1.2;
T=10;
w=2*pi/T;
cd=1;
cm=2;
rho=1025;
L=1.56*T^2;
k=2*pi/L;
t=0:0.1:100;
dt=1
y=0:dt:h
```

```
alpha=30;
z=size(y)
x=y*tand(alpha);
for i=1:z(2)
u(i,:)=(w*H/2)*(cosh(k*y(i))/sinh(k*h))*cos(k*x(i)-w*t);
   v(i,:)=(w*H/2)*(sinh(k*y(i))/sinh(k*h))*sin(k*x(i)-w*t);
   ax(i,:)=(w^2*H/2)*(cosh(k*y(i))/sinh(k*h))*sin(k*x(i)-w*t);
   ay(i,:)=-(w^2*H/2)*(sinh(k*y(i))/sinh(k*h))*cos(k*x(i)-w*t);
   Vn(i,:)=u(i,:)*cos(alpha)+v(i,:)*sin(alpha);
   An(i,:)=ax(i,:)*cos(alpha)+ay(i,:)*sin(alpha);
   Fd(i,:)=0.5*rho*cd*D*abs(Vn(i,:)).*Vn(i,:);
   Fi(i,:)=cm*rho*(pi*D^2/4)*An(i,:);
   FT(i,:)=Fd(i,:)+Fi(i,:);
   M(i,:)=FT(i,:)*y(i);
end
Ft=sum(FT);
MI=sum(M);
plot(t,Ft)
```

Exercise in HSE:

1. What are the main hazards related to the oil and gas industry? Safety and injury hazards, health and illnesses hazards.

2. Write a short note on occupational safety and health management systems. The insinuation of implementing an occupational safety and health management system at all workplaces came into the limelight when the "Global Strategy on Occupational Safety and Health: Conclusions" was adopted by the "International Labour Conference" at its 91st session in 2003. The strategy advocates the application of a systems approach to the management of national OSH systems. Also, guidelines on Occupational Safety and Health Management Systems provide a national/organizational framework for OSH management systems. As per these guidelines, the OSH management system should contain the main elements of policy, organizing, planning and implementation, evaluation, and action for improvement.

3. What key features should an efficient safety and health management system fulfill?
   It should ensure the safety of different operational sites by correctly mapping the business processes, risks, and controls involved in the oil and gas industry's three segments (upstream, midstream, and downstream). It should enable workers to follow consistent health and safety practices. It should help manage site inspections, permits, violations, lessons learned, and best practices execution for the oil and gas sector. It must be well documented (strategies and action plans) and quickly understood and readily available to all the workers.

4. Name some components of an adequate occupational safety and health management system.

Health and safety plan, administration, work area management, H&S risk management, inventory management, etc.

5. What are the benefits of occupational safety and health management systems?

It enables the oil and gas industry to perform hazard identification, risk assessment, and implementation of various control methods; it ensures the well-being of all the employees, and thus, contributes to a more inspired and performance-driven workforce. Regular risk assessment process helps in frequent tracking and monitoring of health and safety indicators (both leading and lagging). Reduced costs associated with accidents and incidents improved regulatory compliance implementation of the OSH management system, giving a competitive edge and improving relationships between stakeholders, such as clients, contractors, subcontractors, con consultants, suppliers, employees, and unions.

6. What are the different safety measures in design and process operation employed in oil and gas industries?

Inerting, explosion, fire prevention, and sprinkler systems.

7. What are the different conditions that must be satisfied to cause fire accidents?

Presence of combustive or explosive material, presence of oxygen to support combustion reaction, source of ignition to initiate the reaction.

8. What are the different fire and explosion control measures?

Use explosion-proof equipment and instruments, use well-designed sprinkler systems, and use modern design features.

9. What is the significance of inerting and purging methods?

Reduce oxygen or fuel concentration below the target value. Usually, it is 4% below the limiting oxygen concentration; nitrogen, carbon dioxide, and others can be used; nitrogen is commonly used.

10. What are the different purging methods available in process industries?

Vacuum purging, pressure purging, combined purging, vacuum and pressure purging with impure nitrogen, sweep-through purging, and siphon purging.

11. How are flammability diagrams helpful in reducing fire hazards?

It determines whether or not an explosive mixture exists and provides a target concentration for inerting and purging; two distinct uses are placing the vessel out of service and putting it into service.

12. What is meant by placing a vessel out of service and placing a vessel into service?

Placing vessel out of service: Gas concentration at points R and M are known from the flammability diagram of the fuel (e.g., methane); find the composition at point S graphically placing vessel into service: gas concentration at points R and M are known, composition at point S can be determined graphically, nitrogen is pumped in till the point S is reached.

13. What are the different types of sprinkler systems?
    Antifreeze sprinkler system, deluge sprinkler system, dry pipe sprinkler system, wet pipe sprinkler system.

14. Discuss the ventilation guidelines used inside the storage area.
    The system should be interlocked with a sound alarm when ventilation fails; inlet and exhausts should be located to provide air movement across the entire area; and recirculation of air is permitted, stopped when air concentration > 25% of the lower flammability limit.

15. What are the significant problems and risks involved in drilling operations?
    Drill pipe sticking and pipe failure, lost circulation, hole deviation and borehole instability, mud contamination, formation damage, and drill bit failure.

16. Name some quantitative risk analysis software used in the process industries.
    Safety, Phast risk, Risk, Risk spectrum, ASAP, Plato.

17. Name some of the standards used for atmospheric storage tanks. Discuss anyone in detail.
    API 650, API 620, ASME sec V, ASTM, etc.

18. What are the factors which make the management systems rigorous?
    Complexity, hazard and risk, resource demands/availability, culture.

19. Write a short note on process hazard analysis.
    Process hazard analysis is a thorough, orderly, systematic approach to identifying, evaluating, and controlling the hazards of processes involving highly hazardous chemicals. The employer must perform an initial process hazard analysis (hazard evaluation) on all processes covered by this standard. The process hazard analysis methodology selected must be appropriate to the complexity of the process and must identify, evaluate, and control the hazards involved in the process.

20. Differentiate between OSHA's process safety management regulations and EPA's risk management program.
    PSM: Protects the workforce, protects contractors, protects visitors to the facility, basically protects the workplace
    RMP: Protects the community, protects the general public around the facility, and protects adjacent facilities such as schools and hospitals.

1. The relationship between the frequency and number of people suffering a given level of harm from the realization of hazard is called ....
   Societal risk.

2. Estimating uncertainties associated with the entire risk assessment process is called ....
   Risk characterization.

3. .... can be a suitable tool for evaluating industrial fire risk and prioritizing units at the general level of an industrial complex, especially chemicals company.
   Frank and Morgan risk analysis.

4. The control score for a department in an oil and gas industry is given as 156, and the hazard score is 152. Calculate the percentage risk index.
   (a) 24.04   (b) 26.02   (c) –26.02   (d) –24.04
   (e) –24.04

5. Action taken to control or reduce risk is called .... risk aversion.

6. In a risk assessment context, what do you understand by the term risk?
   (a) An unsafe act or condition.
   (b) Something with the potential to cause injury.
   (c) Any work activity that can be described as dangerous.
   (d) The likelihood that harm from a particular hazard will occur.
   (d) The likelihood that harm from a particular hazard will occur.

7. .... are used for representing societal risk.
   FN curves.

8. Prevention of hazard occurrence through proper hazard identification, assessment, and elimination is called ....
   Safety.

9. The occurrence of a single or sequence of events that produce unintended loss is called ....
   Accident.

10. Chemical or physical condition that can potentially cause damage to people, property, or environment is called ....
    Hazard.

11. The measure of expected effects of the results of an incident is called:
    (a) hazard   (b) consequence   (c) failure   (d) incident
    Answer: (b) consequence.

12. Define individual risk and societal risk.
    Individual risk: Defined as the frequency at which individuals may be expected to sustain a given level of harm from realizing the hazard. It usually accounts for only the risk of death. It is expressed as risk per year.
    Societal risk: Defined as a relationship between the frequency and number of people suffering a given level of harm from the realization of a hazard.

Societal risks are expressed as FN curves, showing the relationship between the cumulative frequency (F) and the number of fatalities (N).

13. What is the difference between safety and risk?

Safety or loss prevention: preventing hazard occurrence (accidents) through proper hazard identification, assessment, and elimination.

Risk: a measure of the magnitude of damage along with its probability of occurrence.

14. What are the application issues of risk assessment?

Risk assessment often relies on inadequate scientific information or lack of data. For example, any data related to repair may not be helpful to assess newly designed equipment. It means that even though there is less data available, all data related to that event cannot be considered qualified data for risk assessment.

15. State a few golden rules of a good HSE Management program. Identifies and eliminates existing safety hazards.

Safety knowledge, safety experience, technical competence, safety management support, commitment to safety.

16. What do you understand by loss? What do you understand by acceptable risk? As an employee of an oil industry, how do you react to the term acceptable risk?

Loss: Severity of negative impact.

Acceptable risk: Level of human and/or material injury or loss from an industrial process considered tolerable by a society or authorities given the social, political, and economic cost-benefit analysis.

17. Explain about safety assurance and safety assessment methods.

Safety assurance: This is the application of safety engineering practices intended to minimize the risks of operational hazards. Strategies include reactive, proactive, predictive, and iterative. Risk analysis is one of the methods.

Safety assessment: Assessed to their potential severity of impact (generally a negative effect, such as damage or loss) and the probability of occurrence. Methods: risk assessment, hazard identification, risk characterization, etc.

18. What are goal-setting regimes and rule-based regimes?

Goal-setting regimes: Dutyholder assesses risk. Should demonstrate its understanding, controls cover management, technical, and systems issues. Keeps pace with new knowledge opportunity for workforce engagement

Rule-based regimes: The legislator sets the rules. Emphasizes compliance rather than outcomes. Slow to respond. Less emphasis on continuous improvement. Less workforce involvement.

19. Explain the importance of safety in HSE management through a schematic illustration.

Importance of safety......

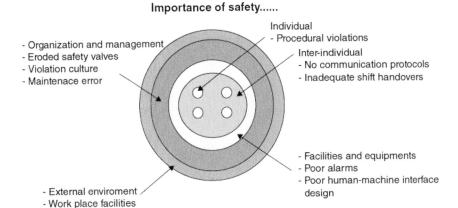

- Organization and management
- Eroded safety valves
- Violation culture
- Maintenace error

Individual
- Procedural violations

Inter-individual
- No communication protocols
- Inadequate shift handovers

- Facilities and equipments
- Poor alarms
- Poor human-machine interface design

- External enviroment
- Work place facilities

20. Calculate the risk ranking for each department.

Exposure dept	Hazard score	Control score	Property value ($\times 10^3$)	Business interruption cost ($\times 10^3$)	Composite score	
					Personnel	Exposure (dollars)
A	257	304	2900	1400	900	5200
B	71	239	890	1200	653	2743
C	181	180	1700	720	1610	4030
D	152	156	290'	418	642	1350
E	156	142	520	890	460	1870
F	113	336	2910	3100	1860	7870

A-1, C-2, E-3, D-4, B-5, F-6.

21. The influx of fluids from the formation into the wellbore is called
....
Well kick.

22. Offshore reserve that can't economically support installing fixed drilling and production platforms is called ... ...
Marginal field.

23. What are the challenges in offshore drilling?
(a) Complex operations involve (a) innovative equipment, (b) skilled labor, (c) all of the above.
Answer: (c) complex operations, innovative equipment, skilled labor.

24. The influx of fluid from the formations into the wellbore is called:
(a) dispersion, (b) diffusion, (c) well kick, (d) blowout.
Answer: (a) dispersion.

25. ... . maintain control over a potential high-pressure condition that exists in the formation.
BOP

26. What are the critical factors in drilling from a safety point of view?
    System design is the "complete integration of all parts into the whole, which should be considered in the beginning itself." Consultations are required between field development engineers, equipment manufacturers, service engineers, maintenance engineers, drilling companies, reservoir engineers, etc.

27. List different problems associated with offshore drilling operations. Also, comment on the recent development of alternate drilling techniques to improve safety in operations.
    - Highly complex and technically challenging operation.
    - Uses innovative equipment and techniques.
    - Require highly special individuals to design/execute the drilling operation.

28. Three systems are commonly used as a measure of accident. What are they? Name them. Also, indicate the most important common feature between them.
    - OSHA (Occupational Safety and Health Administration, US Dept of Labor).
    - Fatal Accident Rate (FAR).
    - Fatality rate or deaths per person per year.
    - All three methods report the number of accidents and/or fatalities for a fixed number of working hours during a specified period.

29. What are the steps taken to defeat an accident process? List different types of risk as identified in risk analysis studies.
    Different types of risk include strategic, financial, compliance, and operations.

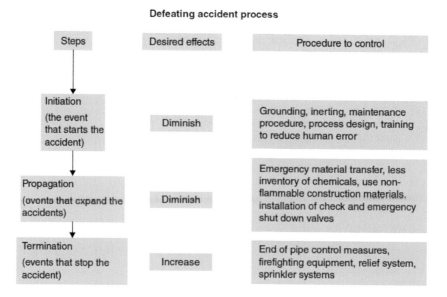

Defeating accident process

Steps	Desired effects	Procedure to control
Initiation (the event that starts the accident)	Diminish	Grounding, inerting, maintenance procedure, process design, training to reduce human error
Propagation (events that expand the accidents)	Diminish	Emergency material transfer, less inventory of chemicals, use non-flammable construction materials, installation of check and emergency shut down valves
Termination (events that stop the accident)	Increase	End of pipe control measures, firefighting equipment, relief system, sprinkler systems

30. What are the advantages and disadvantages of using leg drilling?
    Advantages:
    - Early production for improved cash flow.
    - Several wells in a leg can be completed and placed in production.
    - The drilling rig moves to a thriving cluster in another leg.
    - When wells in the 2nd leg are drilled and completed, they can be placed in production.
    - Continuous flow is maintained.
    - Time and money savings if two rigs are used.
    - Use a standard rig for drilling and a lighter rig for completion works.
    - While the completion rig completes the work while drilling proceeds in another leg well cluster.
    - Elapsed time can be reduced.
    - Cost savings due to reduced on-site requirement of heavier drilling rigs.

Disadvantages:
    - Limited to the size of the completion equipment used.
    - Major limitation.
    - Number of wells that can be practically installed in a given leg.

31. The first step in all risk assessment or QRA study.
    HAZARD.

32. If the hazard evaluation shows low probability and minimum consequence, then the system is called . . . .
    Gold plated.

33. . . . . identifies potential hazards and operability problems due to deviations.

HAZOP.
34. . . . . is a logical, structured process that can help identify potential causes of system failure, such as causes of initiating events or failure of barrier systems.

FTA.
35. . . . . is the most commonly used probabilistic analysis method used for hazard identification.

FTA.
36. Which one of them is a primary keyword?
    (a) more, (b) reverse, (c) erode, (d) fluctuation.
    Answer: (c) erode.

37. What is a HAZARD?
    (a) Where is an accident likely to happen? (a) an accident waiting to happen, (b) something with the potential to cause, (c) the likelihood of something going wrong.
    Answer: (b) something with the potential to cause.

38. What are the different hazard identification methods? Explain them briefly.
    - Process hazard checklists.
    - Hazard surveys.
    - HAZOP.
    - Safety review.

39. Explain about hazard control, hazard evaluation, and hazard monitoring.

    Hazard control: Sometimes, hazards can be eliminated, but most often, measures have to be put in place to manage hazards efficiently, and it also helps to be systematic. This step-by-step procedure starts from the big ones, like whether to repair or upgrade the equipment and works down until you find a practical solution.

    Hazard evaluation: Hazard evaluation can be performed at any stage. The system is called gold-plated if the hazard evaluation shows a low probability and minimum consequence. Potentially unnecessary and expensive safety equipment and procedures are implemented in the system.

    Hazard monitoring: Hazard controls must be reviewed periodically to ensure they are still effective and appropriate. This can be part of your regular safety inspections. If you have one, talking with staff and the Joint Health and Safety Committee (if you have one) is an excellent way to start to get an idea of how healthy controls are working and what could be done even better. Some questions to consider when reviewing hazard controls are:

    - Is the hazard under control?
    - Have the steps taken to manage it solved the problem?
    - Are the risks associated with the hazard under control too?
    - Have any new hazards been created?

40. What is meant by hazard analysis?
    - Identification of undesired events that led to the materialization of a hazard.
    - Analysis of the mechanisms by which these undesired events could occur.
    - Estimation of the extent, magnitude, and likelihood of any harmful effects.

41. .... is a rating corresponding to the seriousness of an effect of a potential failure.
    Severity.

42. The objective of FMEA is on .... and not on............ .... Failure prevention and detection.

43. Write short notes on HAZID and its limitations (if any).
    - Deals with engineering failure assessment.
    - Evaluate the reliability of specific segments of a plant operation.
    - To determine probabilistic results of failure.

- Fault tree analysis is one such common form of engineering failure assessment.
- Limitations: It is not identified until an accident occurs.

44. Name one hazard evaluation method used for mechanical and electrical systems.

FMEA.

45. What do you understand by a weak link? This is required to be identified in what kind of hazard studies?
   - Weak link will be the one that has the highest rank of failure.
   - Do a detailed analysis of the components present in the weak link.
   - One may also re-design to reduce the components' failure probability in the weak link.

This is identified while conducting FMEA.

46. Name two types of FMEA
   (a) design FMEA, (b) process FMEA.

47. What advantages does HAZOP have when applied to a new design?
   - HAZOP supplements the design ideas with imaginative anticipation of deviations. These may be due to equipment malfunction or operation error.
   - In the design of new plants, designers overlook a few issues related to safety in the beginning. HAZOP highlights these errors.
   - HAZOP is an opportunity to correct these errors before such changes become too expensive or impossible. HAZOP methodology is widely used to aid in loss prevention.
   - HAZOP is a preferred tool for risk evaluation

48. Draw an FMEA cause and effect diagram for an airbag used in a passenger car.

FMEA cause and effect diagram
Example 2 — air bag in passenger car

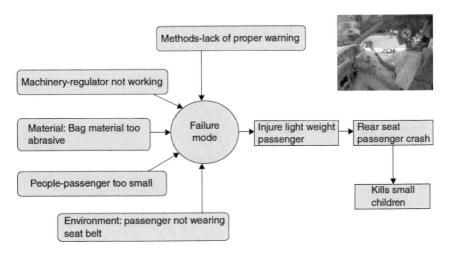

49. Explain full recording and recording by exception.

   Full recording: Later practices were to report everything. Each keyword is clearly stated as applied to the system under study. Even statements like "no cause could be identified" or "no consequence arose from the cause recorded" are seen in these statements.

   Recording by exception: Only potential deviations with some negative consequences were recorded in earlier HAZOP reports. Also, for hand-written records, it certainly reduces the time—both in the study itself and subsequent production of the HAZOP report. In this method, it is assumed that anything that is not included is deemed to be satisfactory.

50. Conduct FMEA analysis for the anti-skid braking system.

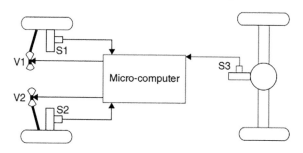

The figure shows the layout plan of the passenger car's anti-skid braking system. The objective is to prevent the locking of the front wheels during heavy braking under bad road conditions. Speed sensors S1 and S2 measure the speed of two front wheels. S3 measures the speed of the drive shaft. This also indicates the speed of the rear wheel. Signals from three-speed sensors are fed to a microcomputer. If the speed of the front wheels falls significantly low, meaning the application of brakes, then valves V1 and V2 are opened to reduce the braking force.

The FMEA-anti-skid braking system of car

Practice questions:

1. Identify significant ways to prevent accidents resulting from fire and explosions.
2. Three systems are commonly used as a measure of accidents. What are they? Name them. Also, indicate the most important common feature between them.
3. Define individual risk and societal risk.
4. What do you understand by acceptable risk? As an employee of an oil industry, how do you react to the term acceptable risk?

TABLE 7.1
Failure Mode Effect analysis worksheet

Component	Failure Mode	Failure Effect	Comment
Front-wheel sensor S1, S2	No output signal	The computer will assume that one wheel has stopped.	Uneven braking on front wheels
		Sends a signal to open the relief valve on that wheel.	The alarm system required to switch off the computer
		Results in partial loss of front wheel braking	
	Fail to open	One front wheel could lock on heavy braking	Not desired. Test facility required
Front-wheel valves V1, V2	Fail to close	Partial loss of front brake	Uneven braking on front wheels Additional stop Valve required?
Rear wheel sensor, S3	No output signal	The microcomputer will have no reference speed from the rear wheel	Alarm system required
		Will not attempt to close V1 or V2	
		Both front wheels could lock on heavy braking	
	No output signals to either front wheel valves	Both front wheels could lock on heavy braking	Alarm system required
Microcomputer	No output signal to one front wheel valve	One front wheel could lock on heavy braking	Alarm system required
	Spurious output to both front-wheel valves	Total loss of front wheel braking	The alarm system required to switch off the computer
	Spurious output to one front wheel valve	Partial loss of front wheel braking	The alarm system required to switch off the computer

5. You are given two options to reach Station A from Station B.
   (a) You wish to drive the complete distance of 2200 km at an average speed of 45 km/h to reach Station A by road; (b) alternatively, you plan to fly and reach Station B by commercial airlines in 2½ h.

Answer the following questions:

1. Which travel is the safest, based on the FAR in general? Explain. Refer to the table for fatalities of different modes of transport.
2. What is the fatality rate for the safest trip?
3. Suppose you travel by car at an average 60 km/h speed. Do you think FAR will change? Will it increase or decrease? Guess the answer to this question based on calculations done for the previous questions.

Justify your answer without working out the FAR in detail.

TABLE 7.2
FAR for common activities

Activity	FAR (deaths/10^8 h)
Staying at home	4
Traveling by car	57
Bicycle riding	96
Traveling by air	240
Motorcycle riding	660
Rock climbing	4000

TABLE 7.3
FAR for Industry

Industry	FAR
Chemical industry	2
Factory work	4
Coal mining	8
sea fishing	40
Offshore oil and gas	62
Steel fabricators	70

6. An employee works in a process industry with a FAR of 4. This industry has regular working hours. As the employee gained experience in his
   trade, he wishes to change his job. Another oil and gas company abroad
   offered him a job. The work agreement of the new company says that his
   working hours are only 4 hours per shift and he will have to work only for
   200 days a year.

   - For your reference, see the table showing the FAR for different
     industries.
   - The employee is confused as he foresees a higher risk rate in the oil
     and gas industry compared to the current process industry where he is
     employed. But he expects a good financial gain.

Answer the following:

- Should the employee opt for a change in his job? Being an HSE consultant,
  should you advise him to do so, explain the basis on which you will work
  out his safety in the new job.
- Suppose the employee wants to shift back to his original employer after
  his abroad assignment is over, should you advise him to bargain toward his
  working hours so that he faces the same fatality rate as that of his recent
  abroad assignment? If so, state briefly your advice to him.

References

Abbasi T, Abbasi SA (2007) Boiling liquid expanding vapor explosion (BLEVE): Mechanism, consequence assessment, and management. *Journal of Hazardous Materials*, 141:489–519.

ABS (2004) *Offshore LNG terminals*. American Bureau of Shipping, Houston, TX, USA.

ABS (2014) *LNG bunkering: Technical and operational advisory*. American Bureau of Shipping, Houston, TX, USA.

Adams AJ, Baltrop NDP (1991) *Dynamics of fixed marine structures*. Butterworth-Heinemann Ltd., London.

Adrezin R, Bar-Avi P, Benaroya H (1996) Dynamic response of compliant offshore structures-review. *Journal of Aerospace Engineering*, 9(4):114–131.

Adrezin R, Benaroya H (1999) Non-linear stochastic dynamics of tension leg platforms. *Journal of Sound and Vibration,* 220(1):27–65.

Agarwal AK, Jain AK (2002) Dynamic behavior of offshore spar platforms under regular sea waves. *Ocean Engineering*, 30:487–516.

Agrawal A, Wei Y, Holditch SA (2012) A technical and economic study of completion techniques in five emerging US gas shales: A Woodford Shale example. *SPE Drilling & Completion,* 27(01):39–49.

Aktan AE, Catbas FN, Turer A, Zhang ZF (1998a) Structural identification: Analytical aspects. *Journal of Structural Engineering*, 124(7):817–829.

Aktan AE, Helmicki AJ, Hunt VJ (1998b) Issues in health monitoring for intelligent infrastructure. *Smart Materials and Structures*, 7(5):674–692.

Ale BJM (2002) Risk assessment practices in the Netherlands. *Safety Science,* 40:105–126.

Alvarado V, Manrique E (2010) *Enhanced oil recovery: Field planning and development strategies*. Gulf Professional Publishing, Boston.

Amenola A, Continin S, Ziomas I (1992) Uncertainties in chemical risk assessments: Results of European Benchmark exercise. *Journal of Hazardous Materials*, 29:347–363.

American Petroleum Institute (2005) *Design and analysis of station keeping systems for floating structures, API recommended practice 2SK*, 3rd edition. American Petroleum Institute, 200 Massachusets Avenue, Washington, DC.

Anagnostopoulos SA (1982) Dynamic response of offshore structures to extreme waves including fluid-structure interaction. *Engineering Structures*, 4:179–185.

API RP 2T (1997) *Recommended practice for planning, designing, and constructing tension leg platforms,* 2nd edition. American Petroleum Institute, Washington, DC.

API RP WSD (2005) *Recommended practice for planning, designing and constructing fixed offshore platforms-working stress design*. American Petroleum Institute, Washington, DC.

API-RP 2A-WSD (2000) *Recommended practice for planning, designing, and constructing fixed offshore platforms*. American Petroleum Institute, Washington, DC.

Arnott AD, Greated CA, Incecik A, McLeary A (1997) An investigation of extreme wave behavior around the model TLP. *Proceedings of 13th International Offshore and Polar Engineering Conference, ISOPE,* Honolulu, HI, May 25–30, pp. 185–192.

Arshad A (2011) Hazard identification and management in oil and gas industries. M.Tech (Petroleum Engg) dissertation submitted to IIT Madras, India.

Austin EH (2012) *Drilling engineering handbook*. Springer Science & Business Media, Netherlands.

Aven T, Vinnem JE (2007) *Risk management with applications from offshore petroleum industry*. Springer, London, 200 pp.

Bahadori A (2018) *Fundamentals of enhanced oil and gas recovery from conventional and unconventional reservoirs*. Gulf Professional Publishing, Boston.

Bai Y (2001) Pipelines and risers. In: *Elsevier ocean engineering book series,* Vol. 3. Elsevier, Amsterdam.

Baihly JD, Malpani R, Edwards C, Han SY, Kok JC, Tollefsen EM, Wheeler CW (2010) Unlocking the shale mystery: How lateral measurements and well placement impact completions and resultant production. In: *SPE unconventional resources conference/gas technology symposium*, November (pp. SPE-138427). SPE, San Antonio, TX.

Bar-Avi P (1999) Nonlinear dynamic response of a tension leg platform. *Journal of Offshore Mechanics and Arctic Engineering*, 121:219–226.

Bar-Avi P, Benaroya H (1996) Non-linear dynamics of an articulated tower in the ocean. *Journal of Sound and Vibration*, 190(1):77–103.

Basim MB, Johnson C, Philip J, Roesset M (1996) Implication of tendon modeling on the nonlinear response of TLP. *Journal of Structural Engineering*, 122(2):142–149.

Bea RG, Xu T, Stear J, Ramas R (1999) Wave forces on decks of offshore platforms. *Journal of Waterway, Port, Coastal, and Ocean Engineering*, 125(3):136–144.

Bearman PE, Russell MP (1996) Viscous damping of TLP hulls. In: *Managed program on uncertainties and loads on offshore structures (ULOS), project report*. Imperial College, London.

Bearman PW (1984) Vortex shedding from oscillating bluff bodies. *Annual Review of Fluid Mechanics*, 16:195–222.

Bhattacharyya SK, Chandrasekaran S, Prasad R (2010a) Event analysis for offshore riser failure. *Proceedings of the First International Conference on Drilling Technology, ICDT 2010*, IIT Madras, India, November 18–20.

Bhattacharyya SK, Chandrasekaran S, Prasad R (2010b) Risk assessment for off-shore pipelines. *Proceedings of the First International Conference on Drilling Technology, ICDT 2010*, IIT Madras, India, November 18–20.

Blevins RD (1994) *Flow-induced vibration*, 2nd edition. Krieger Publishing, Malabar, FL.

Boaghe OM, Billings SA, Stansby PK (1998) Spectral analysis for non-linear wave forces. *Journal of Applied Ocean Research*, 20:199–212.

Bonvicini S, Leonelli P, Spadoni G (1998) Risk analysis of hazardous materials transportation: Evaluating uncertainty using fuzzy logic. *Journal of Hazardous Materials*, 62:59–74.

Boom WC, Pinkster JA, Tan PSG (1984) Motion and tether force prediction of a TLP. *Journal of Waterway, Port, Coastal, and Ocean Engineering*, 110(4):472–486.

Booton M, Joglekar N, Deb M (1987) The effect of tether damage on tension leg platform dynamics. *Journal of Offshore Mechanics Arctic Engineering*, 109:186–192.

Bottelberghs PH (2000) Risk analysis and safety policy developments in the Netherlands. *Journal of Hazardous Materials*, 71:59–84.

Brazier AM, Greenwood RL (1998) Geographic information systems: A consistent approach to land use planning decisions around Hazardous installations. *Journal of Hazardous Materials*, 61:355–361.

Brika D, Laneville A (1993) Vortex-induced vibrations of a long flexible circular cylinder. *Journal of Fluid Mechanics*, 250:481–508.

Bringham EO (1974) *The fast Fourier transform*. Prentice Hall, Englewood Cliffs, NJ.

Brode HL (1959) Blast wave from a spherical charge. *Physics of Fluids*, 2(2):217–229.

BS6235 (1982) *Code of practice for fixed offshore structures*. British Standards Institution, London.

Bubbico R, Marchini M (2008) Assessment of an explosive LPG release accident: A case study. *Journal of Hazardous Materials*, 155(3):558–565.

Buchner B, Bunnik T (2007) Extreme wave effects on deepwater floating structures. *Proceedings of Offshore Technology Conference, OTC 18493*, Houston, TX, April 30–May 3.

Buchner B, Wichers JEW, de Wilde JJ (1999) Features of the state-of-the-art deepwater offshore basin. *Proceedings of Offshore Technology Conference, OTC 10841*, Houston, TX, May 3–6.

Burrows R, Tickell RG, Hames D, Najafian G (1992) Morison wave forces co-efficient for application to random seas. *Journal of Applied Ocean Research*, 19:183–199.

Cairns WJ (Ed) (1992) *North Sea oil and the environment: Development oil and gas resources, environmental impacts and responses*. International Council of Oil and the Environment, Elsevier, London.

Cano ML, Dorn PB (1996) Sorption of two model alcohol ethoxylate surfactants to sediments. *Chemosphere*, 33:981–994.

Capanoglu CC, Shaver CB, Hirayama H, Sao K (2002) Comparison of model test results and analytical motion analyses for buoyant leg structure. *Proceedings of 12th International Offshore and Polar Engineering Conference*, Kitakushu, Japan, May 26–31, pp. 46–53.

Chakrabarti SK (1971) Nondeterministic analysis of offshore structures. *Journal of Engineering Mechanics Division*, 97.

Chakrabarti SK (1980) Inline forces on the fixed vertical cylinder in waves. *Journal of Waterway, Port, Coastal and Ocean Engineering*, 106:145–155.

Chakrabarti SK (1984) Steady drift force on vertical cylinder—viscous versus potential. *Applied Ocean Research*, 6(2):73–82.

Chakrabarti SK (1987) *Hydrodynamics of offshore structures: Computational mechanics*. WIT Press, Southampton.

Chakrabarti SK (1990) *Non-linear method in offshore engineering*. Elsevier Science Publisher, the Netherlands.

Chakrabarti SK (1994) *Offshore structure modeling*. World Scientific, Singapore.

Chakrabarti SK (1998) Physical model testing of floating offshore structures. *Dynamic Positioning Conference, Session Index, Design*, Houston, TX, October 13–14.

Chakrabarti SK (2002) *The theory and practice of hydrodynamics and vibration*. World Scientific, Singapore.

Chakrabarti SK (2005) *Handbook of offshore engineering*, Vol. II. Offshore Structure Analysis, Plainfield, IL.

Chakarabarti SK, Hanna SY (1990) Added mass and damping of a TLP column model. *Proceedings of the International Offshore Technology Conference No. 4643*, Houston, TX, pp. 533–545.

Chakrabarti SK, Tam WA (1975) Wave height distribution around the vertical cylinder. *Journal of Waterways Harbors and Coastal Engineering Division*, 101:225–230.

Chakrabarti SK, Wolbert AL, Tam AW (1976) Wave force on the vertical cylinder. *Journal of Waterways, Harbors, and Coastal Engineering Division*, 10(2):203–221.

Chamberlain GA (1987) Development in design methods for predicting thermal radiation from flares. *Chemical Engineering Research and Design*, 65, 299–309.

Chandrasekaran S (2007) Influence of wave approach angle on TLP's response. *Ocean Engineering*, 34:1322–1327.

Chandrasekaran S (2010a) Chemical risks—an overview. *Keynote address at HSE in Oil and gas exploration and production, International HSE Meet*, IBC-Asia, Kuala Lumpur, Malaysia, December 6–8.

Chandrasekaran S (2010b) Risk assessment of offshore pipelines. *Keynote address at HSE in Oil and gas-exploration and production, International HSE Meet*, IBC-Asia, Kuala Lumpur, Malaysia, December 6–8.

Chandrasekaran S (2011a) Hazard identification and management in oil and gas industry using Hazop. *Proceedings of the Seminar on Human Resource Development for Offshore and Plant Engineering (HOPE)*, Changwon University, South Korea, April 2011, pp. 1–10.

Chandrasekaran S (2011b) Health, safety and environmental management in petroleum and offshore engineering. *Proceedings of the Seminar on Human Resource Development for Offshore and Plant Engineering (HOPE),* Changwon University, South Korea, April 2011, pp. 23–28.

Chandrasekaran S (2011c) Quantitative risk assessment of Group Gathering Station (GGS) of oil exploration and production. *Proceedings of the Seminar on Human Resource Development for Offshore and Plant Engineering (HOPE),* Changwon University, South Korea, April 2011, pp. 11–22.

Chandrasekaran S (2011d) Strategic rig project commissioning and risk management. *Keynote address at Post-Conference Workshop on International Conference on Offshore Drilling Rigs,* IBC Asia, Singapore, July 24–29.

Chandrasekaran S (2011e) Risk assessment and management in offshore and petroleum industries. *Keynote address at Pre-Conference workshop on International Conference on Asia Pacific HSE Forum on Oil, Gas and Petrochemicals, Fleming Gulf Conferences,* Kuala Lumpur, Malaysia, September 22–24, p. 81.

Chandrasekaran S (2013a) *Dynamic of ocean structures.* Web-based course on the National Program on Technology-enhanced Learning, IIT Madras. Available at: nptel.ac.in/courses/114106036

Chandrasekaran S (2013b) *Ocean structures and materials.* Web-based course on the National Program on Technology-enhanced Learning, IIT Madras. Available at: nptel.ac.in/courses/114106035

Chandrasekaran S (2013c) *Advanced marine structures.* Web-based course on the National Program on Technology-enhanced Learning, IIT Madras. Available at: nptel.ac.in/courses/114106037

Chandrasekaran S (2014a) *Advanced theory on offshore plant FEED engineering.* Changwon National University Press, Republic of South Korea, p. 237. ISBN: 978-89-969792-8-9

Chandrasekaran S (2014b) Heath, safety and environmental management (HSE). *Keynote address at National Research Foundation of Korea, HRD Team for Offshore Plant FEED Engineering, Changwon* National University, South Korea, February 25.

Chandrasekaran S (2014c) Technological advancements in process safety management. *Keynote address in the 4th Annual HSE Excellence Forum in Oil, Gas, and Petrochemicals,* Kuala Lumpur, Malaysia, August 19–21.

Chandrasekaran S (2015a) *HSE in offshore and petroleum engineering.* Lecture notes of online web course, Mass Open-source Online Courses (MOOC), National Program on Technology Enhancement and Learning (NPTEL), Govt. of India.

Chandrasekaran S (2015b) *Dynamic analysis and design of ocean structures,* 1st edition. Springer, New Delhi. ISBN 978-81-322-2276-7

Chandrasekaran S (2015c) *Advanced marine structures.* CRC Press, Boca Raton, FL. ISBN: 9781498739689

Chandrasekaran S (2015d) *Dynamic analysis of offshore structures.* Video course on MOOC, NPTEL portal available at: https://onlinecourses.nptel.ac.in/noc15_oe01/preview

Chandrasekaran S (2016a) *Offshore structural engineering: Reliability and risk assessment.* CRC Press, Boca Raton, FL. ISBN: 978-14-987-6519-0

Chandrasekaran S (2016b) *Risk and reliability of offshore structures.* Video course on MOOC, NPTEL portal available at: https://onlinecourses.nptel.ac.in/noc16_oe01/preview

Chandrasekaran S, Babu R, Ayub A (2010) Hazop study for crude oil pipeline. *Proceedings of 1st International Conference on Drilling Technology, ICDT 2010,* IIT Madras, India, November 18–20.

Chandrasekaran S, Bhaskar K, Harilal L, Brijit R (2010a) Dynamic response behavior of multi-legged articulated tower with and without TMD. *Proceedings of the International Conference of Marine Technology,* Dhaka, Bangladesh, December 11–12, pp. 131–136.

Chandrasekaran S, Bhaskar K, Hashim M (2010b) Experimental study on dynamic response behavior of multi-legged articulated tower. *Proceedings of the 29th International Conference on Ocean, Offshore and Arctic Engineering, OMAE 2010,* Shanghai, China, June 6–11.

Chandrasekaran S, Bhattacharyya SK (2011) *Analysis and design of offshore structures.* HRD Center for Offshore & Plant Engineering (HOPE), Changwon National University, Republic of Korea, p. 285.

Chandrasekaran S, Chandak NR, Gupta A (2006a) Stability analysis of TLP tethers. *Ocean Engineering,* 33(3):471–482.

Chandrasekaran S, Gaurav S (2008) Offshore triangular tension leg platform earthquake motion analysis under distinctly high sea waves. *Journal Ships Offshore Structures,* 3(3):173–184.

Chandrasekaran S, Jain AK (2002a) Dynamic behavior of square and triangular offshore tension leg platforms under regular wave loads. *Ocean Engineering,* 29(3):279–313.

Chandrasekaran S, Jain AK (2002b) Triangular configuration tension leg platform behavior under random sea wave loads. *Ocean Engineering,* 29(15):1895–1928.

Chandrasekaran S, Jain AK (2004) Aerodynamic behavior of offshore triangular tension leg platforms. *Proceedings of ISOPE,* Toulon, France, pp. 564–569.

Chandrasekaran S, Jain AK (2016) *Ocean structures: Construction, materials and operations.* CRC Press, Boca Raton, FL. ISBN: 978-14-987-9742-9.

Chandrasekaran S, Jain AK, Chandak NR (2004) Influence of hydrodynamic coefficients in the response behavior of triangular TLPs in regular waves. *Ocean Engineering,* 31:2319–2342.

Chandrasekaran S, Jain AK, Chandak NR (2006b) Seismic analysis of offshore triangular tension leg platforms. *International Journal of Structural Stability and Dynamics,* 6(1):1–24.

Chandrasekaran S, Jain AK, Chandak NR (2007a) Response behavior of triangular tension leg platforms under regular waves using Stokes nonlinear wave theory. *Journal of Waterway, Port, Coastal, and Ocean Engineering,* ASCE, 133(3):230–237.

Chandrasekaran S, Jain AK, Gupta A (2007b) Influence of wave approach angle on TLP's response. *Ocean Engineering,* 8–9(34):1322–1327.

Chandrasekaran S, Jain AK, Gupta A, Srivastava A (2007c) Response behavior of triangular tension leg platforms under impact loading. *Ocean Engineering,* 34:45–53.

Chandrasekaran S, Jain AK, Madhuri S (2013) Aerodynamic response of offshore triceratops. *Ship Offshore Structures,* 8(2):123–140. doi:10.1080/17445302.2012.691271

Chandrasekaran S, Jamshed N (2015) Springing and ringing response of offshore triceratops. *Proceedings of 34th International Conference on Ocean, Offshore, and Arctic Engineering, OMAE 2015,* St. John's, NL, Canada, May 31–June 5, OMAE2015-41551.

Chandrasekaran S, Kiran A (2014a) Accident Modeling & Risk Assessment of Oil & Gas Industries. *Proceedings of 9th Structural Engineering Convention (SEC 2014),* IIT Delhi, India, December 22–24.

Chandrasekaran S, Kiran A (2014b) Consequence analysis and risk assessment of oil and gas industries. *Proceedings of International Conference on Safety & Reliability of Ship, Offshore and Subsea Structures,* Glasgow, UK, August 18–20.

Chandrasekaran S, Kiran A (2015) Quantified risk assessment of LPG filling station. *Professional Safety, Journal of American Society of Safety Engineers (ASSE),* September 2015, pp. 44–51.

Chandrasekaran S, Lognath RS, Jain A (2015a) Dynamic analysis of buoyant leg storage and regasification platform under regular waves. *Proceedings of 34th International Conference on Ocean, Offshore, and Arctic Engineering, OMAE 2015,* St. John's, NL, Canada, May 31–June 5, OMAE2015-41554.

Chandrasekaran S, Madhavi N (2014a) Retrofitting of offshore structural member using perforated cylinders. *SFA Newsletter*, 13:10–11.

Chandrasekaran S, Madhavi N (2014b) Hydrodynamic performance of retrofitted structural member under regular waves. *International Journal of Forensic Engineering-Inderscience*, 2(2):100–121.

Chandrasekaran S, Madhavi N (2014c) Numerical study on geometrical configurations of perforated cylindrical structures. *Journal Perform Constructed Facilities, ASCE* 30(1):04014185. doi:10.1061/(ASCE)CF.1943-5509.0000687

Chandrasekaran S, Madhavi N (2014d) Variation of water particle kinematics with the perforated cylinder under regular waves. *Proceedings of ISOPE 2014,* June 15–20, Busan, South Korea.

Chandrasekaran S, Madhavi N (2015a) Estimation of force reduction on ocean structures with perforated members. *Proceedings of 34th International Conference on Ocean, Offshore, and Arctic Engineering, OMAE 2015,* St. John's, NL, Canada, May 31–June 5, OMAE2015-41153.

Chandrasekaran S, Madhavi N (2015b) Flow field around an outer perforated circular cylinder under regular waves: Numerical study. *International Journal of Marine Systems and Ocean Technology.* doi:10.1007/s40868-015-0008-1

Chandrasekaran S, Madhavi N (2015c) Design aids for offshore structures with perforated members. *Ship and Offshore Structures,* 10(2):183–203. doi:10.1080/17445302.2014.918309.

Chandrasekaran S, Madhavi N (2015d) Retrofitting of offshore cylindrical structures with different geometrical configurations of perforated outer cover. *International Journal of Shipbuilding Progress*, 62(1–2):43–56. doi:10.3233/ISP-150115

Chandrasekaran S, Madhavi N (2015e) Variation of the flow field around twin cylinders with and without outer perforated cylinder: Numerical studies. *China Ocean Engineering*, 30(5):763–771

Chandrasekaran S, Madhavi N, Natarajan C (2014) Variations of hydrodynamic characteristics with the perforated cylinder. *Proceedings of 33rd International Conference on the Ocean, Offshore and Arctic Engineering, OMAE 2014,* San Francisco, USA, June 8–13.

Chandrasekaran S, Madhuri S (2015) Dynamic response of offshore triceratops: Numerical and experimental investigations. *Ocean Engineering*, 109(15):401–409. doi:10.1016/j.oceaneng.2015.09. 042

Chandrasekaran S, Mayank S, Jain AK (2015b) Dynamic response behavior of stiffened triceratops under regular waves: Experimental investigations. *Proceedings of 34th International Conference on Ocean, Offshore, and Arctic Engineering, OMAE 2015*, St. John's, NL, Canada, May 31–June 5, OMAE2015-41376.

Chandrasekaran S, Merin T (2016) Suppression system for offshore cylinders under vortex-induced vibration. *Vibroengineering Procedia*, 7:01–06.

Chandrasekaran S, Nannaware M (2013) Response analyses of offshore triceratops to seismic activities. *Ships and Offshore Structures*, 9(6):633–642. doi:10.1080/17445302.2013.8 43816

Chandrasekaran S, Pannerselvam R (2009) Offshore structures: Materials, analysis, design and construction. *Proceedings of a Short Course on Offshore Structures and Materials*, IITM, Chennai, December 14–18, p. 156.

Chandrasekaran S, Pannerselvam R, Saravanakumar S (2012) Retrofitting of offshore tension leg platforms with perforated cylinders. *3rd Asian Conference on Mechanics of Functional Materials and Structures (ACFMS),* New Delhi, IIT Delhi, India. December 8–9.

Chandrasekaran S, Parameswara PS (2011) Response behavior of perforated cylinders in regular waves. *Proceedings of 30th International Conference on Ocean, Offshore and Arctic Engineering, OMAE 2011,* Rotterdam, The Netherlands, June 19–24, OMAE 2011-49839.

Chandrasekaran S, Seeram M, Jain AK, Gaurav S (2010) Dynamic response of offshore triceratops under environmental loads. *Proceedings of International Conference of Marine Technology MARTEC,* Dhaka, Bangladesh, December 11–12, pp. 61–66.

Chandrasekaran S, Senger M (2016) Dynamic analyses of stiffened triceratops under regular waves: Experimental investigations. *Ships and Offshore Structures T&F,* http://www.tandfonline.com/doi/full/10.1080/17445302.2016.1200957

Chandrasekaran S, Sharma A, Srivastava S (2007) Offshore triangular TLP behavior using dynamic Morison equation. *Journal Structural Engineering,* 34(4):291–296.

Chandrasekaran S, Sundaravadivelu R, Pannerselvam R, Madhuri S (2011) Experimental investigations of offshore triceratops under regular waves. *Proceedings of 30th International Conference on the Ocean, Offshore and Arctic Engineering,* OMAE 2011, Rotterdam, The Netherlands, June 19–24.

Chandrasekaran S, Yuvraj K (2013) Dynamic analysis of a tension leg platform under extreme waves. *Journal of Naval Architecture and Marine Engineering,* 10:59–68. doi:10.3329/jname.v10i1.14518

Charles WN, Robert CW, Capanoglu C (2005) Triceratops: An effective platform for developing oil and gas fields in deep and ultra-deepwater. *Proceedings of 15th International Offshore and Polar Engineering Conference,* Seoul, Korea, June 19–24, pp. 133–139.

Chaudhury G, Dover W (1995) Fatigue analysis of offshore platforms subjected to sea wave loading. *International Journal of Fatigue,* 7(1):13–19.

Che Hassan CR, Puvaneswaran B, Aziz AR, Noor Zalina M, Hung FC, Sulaiman NM (2009) A case study of consequences analysis of ammonia transportation by rail from Gurun to Port Klang in Malaysia using a safety computer model. *Journal of Safety Health and Environment,* 6(1):Spring, 1–19.

Che Hassan CR, Puvaneswaran B, Aziz AR, Noor Zalina M, Hung FC, Sulaiman NM (2010) Quantitative risk assessment for the transport of ammonia by rail. *American Institute of Chemical Engineers Process Safety Progress,* 29:60–63.

Chen X, Ding Y, Zhang J, Liagre P, Neidzwecki J, Teigen P (2006) Coupled dynamic analysis of a mini TLP: Comparison with measurements. *Ocean Engineering,* 33:93–117.

Choi HS, Lou Jack YK (1991) Nonlinear behavior of an articulated offshore loading platform. *Applied Ocean Research,* 13(2):63–74.

Chong KK, Grieser B, Jaripatke O, Passman A (2010) A completions roadmap to shale-play development: A review of successful approaches toward shale-play stimulation in the last two decades. In: *SPE international oil and gas conference and exhibition in China,* June (pp. SPE-130369). SPE, Beijing.

Chuhan-Jie Z, Bai-quan L, Bing-You J, Qian L, Yi-Du H (2013) Numerical simulation of blast wave oscillation effects on premixed methane air explosions in closed end ducts. *Journal of Loss Prevention in Process Industries,* 29:60–63.

Cipolla CL, Lewis RE, Maxwell SC, Mack MG (2011) Appraising unconventional resource plays: Separating reservoir quality from completion effectiveness. In: *International petroleum technology conference,* November (pp. IPTC-14677). IPTC, Bangkok.

Cipolla CL (2009) Modeling production and evaluating fracture performance in unconventional gas reservoirs. *Journal of Petroleum Technology,* 61(09):84–90.

Clauss GF, Birk L (1996) Hydrodynamic shape optimization of large offshore structures. *Journal of Applied Ocean Research,* 18:157–171.

Clauss GT, Günther C, Eike L, Carsten Ö (1992) *Offshore structures, vol 1—conceptual design and hydromechanics.* Springer, London.

Colwell S, Basu B (2009) Tuned liquid column dampers in offshore wind turbines for structural control. *Engineering Structures,* 31(2):358–368.

Copple RW, Capanoglu CC (1995) A buoyant leg structure for the development of marginal fields in deep water. *Proceedings of Fifth International Offshore and Polar Engineering Conference,* The Hague, The Netherlands, June 16–11, p. 163.

Crawley F, Preston M, Tyler B (2000) *HAZOP: Guide to best practice. guidelines to best practice for the process and chemical industries.* European Process Safety Centre and Institution of Chemical Engineers, Rugby, Warwickshire, UK.

Davenport AG (1961a) The Spectrum of horizontal gustiness near the ground in high winds. *Journal Roy Meteorological Society*, 87:194–211.

Davenport AG (1961b) The application of statistical concepts to the wind loading of structures. *Proceedings Institution of Civil Engineers*, 19:449–471.

David Brown F, William Dunn E (2007) Application of a quantitative risk assessment method to emergency response planning. *Computers & Operations Research*, 34:1243–1265.

Dawson TH (1983) *Offshore structural engineering.* Prentice-Hall, Englewood Cliffs, NJ.

Dean RG, Dalrymple RA (2000) *Water wave mechanics for engineers and scientists, advanced series on ocean engineering*, Vol. 2. World Scientific, Singapore.

Demirbilek Z (1990) Design formulae for offset, set down, and tether loads of a tension leg platform (TLP). *Ocean Engineering*, 17(5):517–523.

Devon R, Jablokow K (2010) Teaching FEED. *Proceedings of Mid-Atlantic ASEE Conference*, Villanova University, Pennsylvania, USA, October 15–16.

DNV (1982) *Rules for the design.* Construction and Inspection of Offshore Structures, Det Norske Veritas, Oslo.

DNV Phast Risk, Det Norske Veritas (2005) User manual-Version 6.7.

DNV-RP-F205 (2010) *Global performance analysis of deep water floating structures.* Det Norske Veritas, Oslo, Norway.

DOE-OG (1985) *Offshore installation: Guidance on design and construction.* U.K. Department of Energy, London.

Donley MG, Spanos PD (1991) Stochastic response of a tension leg platform to viscous drift forces. *Journal of Offshore Mechanics and Arctic Engineering*, 113:148–155.

Dyrbe C, Hansen SV (1997) *Wind loads on structures.* John Wiley and Sons, London.

Dziubinski M, Fratczak M, Markowski AS (2006) Aspects of risk analysis associated with major failures of fuel pipelines. *Journal of Loss Prevention in the Process Industries*, 19:399–400.

El-Gamal AR, Essa, A, Ismail A (2013) Effect of tethers tension force in the behavior of a tension leg platform subjected to hydrodynamic force. *International Journal of Civil, Structural, Construction and Architectural Engineering*, 7(12):645–652.

Engelhard WFJM, de Klepper FH, Hartmann DW (1994) Hazard analysis for the Amoco Netherlands PI1S-PI1S production facilities in the North Sea. *Proceedings of SPE International Conference on Health, Safety and Environment*, Jakarta, January 25–27.

Erik V, Pedro A, Ivan Ø, de Comas FDC (2008) Analysing the risk of LNG carrier operations. *Reliability Engineering and System Safety*, 93:1328–1344. doi:10.1016/j.ress.2007.07.007

Ertas A, Eskwaro-Osire S (1991) Effect of damping and wave parameters on the offshore structure under random excitation. *Nonlinear Dynamics*, 2:119–136.

Ertas A, Lee JH (1989) Stochastic response of tension leg platform to wave and current forces. *Journal Energy Resources Technology*, 111:221–230.

Faltinsen OM (1990) *Sea loads on ships and offshore structures.* Cambridge University Press, New York.

Faltinsen OM, Newman JN, Vinje T (1995) Nonlinear wave loads on a slender vertical cylinder. *Journal of Fluid Mechanics*, 289:179–198.

Finn LD, Maher JV, Gupta H (2003) The cell spar and vortex-induced vibrations. *Offshore Technology Conference, OTC 15244*, Houston, TX, May 5–8, pp. 1–6.

Fjeld, S (1977) Reliability of offshore structures. *Offshore Technology Conference, OTC 3027*, Houston, TX.

Frank KH, Morgan HW (1979) A logical risk process of risk analysis. *Professional Safety*, June, 23–30.

Freudenthal AM, Gaither WS (1969) Design criteria for fixed offshore structures. *Offshore Technology Conference, OTC 1058,* Houston, TX.

Fugazza M, Natale L (1992) Hydraulic design of perforated breakwaters. *Journal of Waterway, Port, Coastal and Ocean Engineering,* 118(1):1–14.

Fujino Y, Abe M (1993) Design formulas for tuned mass dampers based on a perturbation technique. *Earthquake Engineering and Structural Dynamics,* 22(10):833–854.

Furnes, O (1977) Some aspects of reliability of offshore structures, the need for data. *2nd International Conference on Structural Safety and Reliability,* Munich, Germany, pp. 363–366.

Gao C, Du CM (2012) Evaluating the impact of fracture proppant tonnage on well performances in Eagle Ford play using the data of last 3-4 years. In: *SPE annual technical conference and exhibition,* October (pp. SPE-160655). Society of Petroleum Engineers, Houston, TX.

Gasim MA, Kurian VJ, Narayanan SP, Kalaikumar V (2008) Responses of square and triangular TLPs subjected to random waves. *International Conference on Construction and Building Technology, ICCBT 2008,* Universiti Teknologi Petronas, Malaysia, June 16–20.

Gerwick BC Jr (1986) *Construction of offshore structures.* John Wiley, New York.

GESAMP (1991) Global strategies for global environmental protection with addendum. International Maritime Organization, London.

Gie TS, De Boom W (1981) The wave-induced motion of a tension leg platform in deep water. *13th Annual Ocean Technology Conference,* Houston, TX, May 4–7.

Glanville RS, Paulling JR, Halkyard JE, Lehtinen TJ (1991) Analysis of the spar floating, drilling, production, and storage structure. *Offshore Technology Conference, OTC 6701,* Houston, TX, May 6–9, pp. 57–68.

Gomez-Mares M, Zarate L, Casal J (2008) Jet fires and the domino effect. *Fire Safety Journal,* 43, 583–588.

Govardhan R, Williamson CHK (2000) Modes of vortex formation and frequency response of a freely vibrating cylinder. *Journal of Fluid Mechanics,* 420:85–130.

Graff WJ (1981a) *Introduction to offshore structures.* Gulf Publishing Co., Houston.

Graff WJ (1981b) *Introduction to offshore structures: Design. Fabrication and Installation.* Gulf Publishing Co., Tokyo.

Greff K, Greenbauer S, Huebinger K, Goldfaden B (2014) The long-term economic value of curable resin-coated proppant tail-in to prevent flow back and reduce workover cost. In: *SPE/AAPG/SEG unconventional resources technology conference,* August (pp. URTEC-1922860). Society of Petroleum Engineers, Houston, TX.

Guo Boyun, William C, Lyons, Ali Ghalambor (2007) *Petroleum Production Engineering: A Computer-asssited approach,* Gulf Professional Publishing, An imprint of Elsevier, USA, ISBN: 978-0-7506-8270-1.

Guo B, Song S, Chacko J, Ghalamber A (2005) *Offshore pipelines.* Gulf Professional Publishing, MA, USA.

Gurley Kurtis R, Ahsan K (1998) Simulation of ringing in offshore systems under viscous loads. *Journal of Engineering Mechanics* ASCE, 124(5):582–586.

Gusto MSC (2010) The exploration market. *InSide,* 15:4–7.

Halkyard JE, Davies RL, Glanville RS (1991) The tension buoyant tower: A design for deep water. *23rd Annual Offshore Technology Conference, OTC 6700,* Houston, TX, May 6–9, pp. 41–55.

Hari S, Krishna S, Gurrala LN, Singh S, Ranjan N, Vij RK, Shah SN (2021) Impact of reservoir, fracturing fluid and proppant characteristics on proppant crushing and embedment in sandstone formations. *Journal of Natural Gas Science and Engineering,* 95:104187.

Haritos N (1985) Modeling the response of tension leg platforms to the effects of wind using simulated traces. *Mathematics and Computers in Simulation*, 27:231–240.

Haslum HA, Faltinsen OM (1999) Alternative shape of spar platforms for use in hostile areas. *Proceedings of the 31st Offshore Technology Conference, OTC 10953*, Houston, pp. 217–228.

Helvacioglu IH, Incecik A (2004) *Dynamics of double articulated towers, Integrity of offshore structures*, Vol. 4. Elsevier, London.

Henselwood F, Phillips G (2006) A matrix-based risk assessment approach for addressing linear hazards such as pipelines. *Journal of Loss Prevention in the Process Industries*, 19:433–441.

Hitchings GA, Bradshaw H, Labiosa TD (1976) Planning and execution of offshore site investigations for North Sea gravity platform. *Proceedings of Offshore Technology Conference*, Paper No. 2430, Houston, TX.

Hoeg K (1976) Foundation engineering for fixed offshore structures. *Proceedings First International Conference Behaviour of Offshore Structures*, Trondheim, Vol. 1, pp. 39–69.

Hoeg K, Tong WH (1977) Probabilistic considerations in the foundation engineering for offshore structures. *2nd International Conference on Structural Safety and Reliability*, Munich, Germany, pp. 267–296.

Hogben N, Standing RG (1974) Wave loads on large bodies. *Proceedings of the International Symposium on the Dynamics of Marine, Vehicles and Structures in Waves*, University College, London, pp. 258–277.

Hove K, Foss I (1974) Quality assurance for offshore concrete gravity structures. *Proceedings of the Offshore Technology Conference*, Houston, TX, Paper No. 2113, May 6–8.

HSE (2010) *Offshore helideck design guidelines: Health & safety executive*. John Burt Associates, London.

Hsu HT (1981) *Applied offshore structural engineering*. Gulf Publishing Co., Houston.

Humphries JA, Walker DH (1987) Vortex excited response of large-scale cylinders in shear flow. *Proceedings of Sixth OMAE*, Houston, TX, Vol. 2, pp. 139–143.

IEC-61882 (2010) *Hazard and operability studies (HAZOP studies)—application guide*. International Electro Technical Commission, Geneva.

IS1656:2006 (2006) *Indian standard hazard identification and risk analysis-code of practice*. Bureau of Indian Standards, New Delhi.

Isaacson M (1978b) Wave runup around large circular cylinder. *Journal of Waterway, Port, Coastal and Ocean Division*, 104(WW1), 69–79.

Isaacson M, Baldwin J, Allyn N, Cowdell S (2000) Wave interactions with perforated breakwater. *Journal of Waterway, Port, Coastal and Ocean Engineering*, 126(5):229–235.

Isaacson M, Premasiri S, Yang G (1998) Wave interactions with vertical slotted barrier. *Journal of Waterway, Port, Coastal and Ocean Engineering*, 124(3):118–126.

Islam N, Ahmad S (2003) Nonlinear seismic response of articulated offshore tower. *Defense Science Journal*, 53(1):105–113.

Issacson MDetStQ (1982) Non-linear wave effects on fixed and floating bodies. *Journal of Fluid Mechanics*, 120:267–281.

Iwaski H (1981) *Preliminary design study of tension leg platform*. MIT University, Massachussets, USA.

Jain AK (1997) Nonlinear coupled response of offshore TLP to regular waves. *Ocean Engineering*, 24(7):577–592.

Jefferys ER, Patel MH (1982) Dynamic analysis models of tension leg platforms. *Journal of Energy Resources Technology*, 104:217–223.

Jefferys ER, Rainey RCT (1994) *Slender body models of TLP and GBS 'ringing'*. BOSS, McGraw-Hill Inc., New York, NY.

Jin Q, Li X, Sun N, Zhou J, Guan J (2007) Experimental and numerical study on tuned liquid dampers for controlling earthquake response of jacket offshore platform. *Marine Structures*, 20(4):238–254.

Johnson DW, Cornwell JB (2007) Modeling the release, spreading and burning of LNG, LPG, and gasoline on water. *Journal of Hazardous Materials*, 140(3):535–540.

Kakuno S (1983) Reflection and transmission of waves through vertical slit-type structures. *Proceedings of Coastal Structures '83*, ASCE, New York, NY, pp. 939–952.

Kam JCP, Dover WD (1988) Fast fatigue assessment procedure under random time history. *Proceedings of the Institution of Civil Engineers Part 2*, 85:689–700.

Kam JCP, Dover WD (1989) Advanced tool for fast assessment of fatigue under offshore random wave stress history. *Proceedings of the Institution of Civil Engineers Part 2*, 87:539–556.

Kareem A, Datton C (1982) Dynamic effects of wind on TLP. *Proceedings of Offshore Technology Conference, OTC-4229(1)*, Houston, TX, pp. 749–757.

Kareem A, Zhao J (1994) Analysis of non-Gaussian surge response of tension leg platforms under wind loads. *Journal of Offshore Mechanics and Arctic Engineering*, 116:13–144.

Karimirad M, Meissonnier Q, Gao Z, Moan T (2011) Hydroelastic code-to code comparison for a tension leg spar-type floating wind turbine. *Marine Structures*, 24:412–435.

Katbas FN, Aktan AE (2002) Condition and damage assessment: Issues and some promising indices. *Journal of Structural Engineering*, 128(8):1026–1036.

Kawanishi T, Katoh W, Furuta H (1987) Tension leg platform earthquake motion analysis. *Oceans*, 19:543–547.

Kawanishi T, Ohashi S, Takamura H, Kobayashi H (1993) Earthquake response of tension leg platform under unbalanced initial tension. *Proceedings of ISOPE*, Singapore, pp. 319–325.

Kenny FM, James JD, Melling TH (1976) Non-linear wave force analysis of perforated marine structures. *Offshore Technology Conference, OTC 2501*, Houston, TX, pp. 781–796.

Ker W-K, Lee C-P (2002) Interaction of waves and a porous tension leg platform. *Journal of Waterway, Port, Coastal, and Ocean Engineering*, 128(2):88–95.

Khalak A, Williamson CHK (1991) Motions, forces and mode transitions in vortex-induced vibrations at low mass-damping. *Journal Fluids and Structures*, 13:813–851.

Khalifeh M, Saasen A (2020) *Introduction to permanent plug and abandonment of wells* (p. 273). Springer Nature, AG, Cham, Switzerland ISBN: 978-3-030-39969-6, pp. 273.

Khan FI, Abbasi SA (1999) Major accidents in process industries and analysis of causes and consequences. *Journal of Loss Prevention in the Process Industries*, 12(5):361–378.

Kim C-H, Kim MH, Liu YH, Zhao CT (1994) Time domain simulation of the nonlinear response of a coupled TLP system. *International Journal of Offshore and Polar Engineering*, 4(4):281–291.

Kim C-H, Lee C-H, Goo J-S (2007) A dynamic response analysis of tension leg platform including hydrodynamic interaction in regular waves. *Ocean Engineering*, 34(11–12):1680–1689.

Kim C-H, Zhao CT, Zou J, Xu Y (1997) Springing and ringing due to laboratory-generated asymmetric waves. *International Journal of Offshore and Polar Engineering*, 7(1):30–35.

Kim C-H, Zou J (1995) A universal linear system model for kinematics and forces affected by nonlinear irregular waves. *International Journal of Offshore and Polar Engineering*, 5(3):166–170.

King GE (2010) Thirty years of gas shale fracturing: What have we learned? In *SPE annual technical conference and exhibition*, September (pp. SPE-133456). SPE.

King GE (2014) 60 years of multi-fractured vertical, deviated, and horizontal wells: What have we learned? In: *SPE annual technical conference and exhibition*, October (pp. SPE-170952). SPE.

Kiran A (2012) *Risk analyses of offshore drilling rigs*. M.Tech (Petroleum Engg) dissertation submitted to IIT Madras.

Kiran A (2014) *Accident modeling and risk assessment of oil and gas industries*. M.S. (by research) thesis submitted to IIT Madras.

Kjeldsen SP, Myrhaug D (1979) Wave-wave interactions and wave-current interactions in deep water. *Proceedings of 5th POAC Conference Trondheim,* Norway, Vol. 111, p. 179.

Kletz T (2003) *Still going wrong: Case histories and plant disasters.* Elsevier, Chennai, India, pp. 230.

Kobayashi M, Shimada K, Fujihira T (1987) Study on dynamic responses of a TLP in waves. *Journal of Offshore Mechanics Arctic Engineering*, 109:61–66.

Kok J, Moon B, Han SY, Tollefsen E, Baihly J, Malpani R (2010) The significance of accurate well placement in the shale gas plays. In: *SPE unconventional resources conference/gas technology symposium*, November (pp. SPE-138438). Society of Petroleum Engineers, Houston, TX.

Koo BJ, Kim MH, Randall RE (2004) Mathieu instability of a spar platform with mooring and risers. *Ocean Engineering*, 31(17–18):2175–2208.

Kraemer C, Lecerf B, Torres J, Gomez H, Usoltsev D, Rutledge J, . . . Philips C (2014) A novel completion method for sequenced fracturing in the eagle Ford Shale. In: *SPE unconventional resources conference/gas technology symposium*, April (p. D031S007R001). Society of Petroleum Engineers, Houston, TX.

Krishna S, Sreenivasan H, Nair RR (2018) Hydraulic fracture studies of reservoirs with an emphasis on pore fracture geometry studies by developing fracture and microseismic estimations. *Annals of Geophysics*, 61(5):SE554–SE554.

Kurian VJ, Gasim MA, Narayan SP, Kalaikumar V (2008) Parametric study of TLPs subjected to random waves. *International Conference on Construction and Building Technology*, Kuala Lumpur, Malaysia, June 16–20, Vol. 19, pp. 1–9, 213–222.

Kyriakdis I (2003) *"HAZOP—comprehensive guide to HAZOP in CSIRO," CSIRO minerals.* Melbourne, Australia, National Safety Council of Australia.

Laik S (2018) *Offshore petroleum drilling and production.* CRC Press, Baca Raton.

Lee HH, Juang HH (2012) Experimental study on the vibration mitigation of offshore tension leg platform system with UWTLCD. *Smart Structures and Systems*, 9(1):71–104.

Lee HH, Wang P-W (2000) Dynamic behavior of tension-leg platform with net cage system subjected to wave forces. *Ocean Engineering*, 28:179–200.

Lee HH, Wang P-W, Lee C-P (1999) Dragged surge motion of tension leg platforms and strained elastic tethers. *Ocean Engineering*, 26(6):575–594.

Lees FP (1996) Loss *Prevention in process industries: Hazard identification, assessment, and control,* Vol. 1–3. Butterworth-Heinemann, Oxford, 1245 pp.

Leffler WL, Pattarozzi R, Sterling G (2011) *Deep-water petroleum: Exploration and production.* Pennwell Corp., Oklahoma, p. 350.

Leonard JW, Young RA (1985) Coupled response of compliant offshore platforms. *Engineering Structures*, 7:21–31.

Leonelli P, Bonvicini S, Spadoni G (1999) New detailed numerical procedures for calculating risk measures in hazardous material transportation. *Journal of Loss Prevention in the Process Industries*, 12:507–515.

Logan BL, Naylor S, Munkejord T, Nyhgaard C (1996) Atlantic alliance: The next generation tension leg platform. *Proceedings of the Offshore Technology Conference, OTC 8264*, Houston, TX, May 6–9.

Losada IJ, Losada MA, Martin FL (1995) Experimental study of wave-induced flow in a porous structure. *Coastal Engineering*, 26:77–98.

Low YM (2009) Frequency domain analysis of a tension leg platform with statistical linearization of the tendon restoring forces. *Marine Structures*, 22:480–503.

Ma YZ, Holditch S (2015) *Unconventional oil and gas resources handbook: Evaluation and development.* Gulf Professional Publishing, Houston, TX.

Ma YZ, La Pointe PR (Eds.) (2011) *Uncertainty analysis and reservoir modeling: Developing and managing assets in an uncertain world*, AAPG Memoir 96, Vol. 96. AAPG, New York.

Madsen HO, Krenk S, Lind NC (2006) *Methods of structural safety.* Dover Publications, Inc., Mineola, NY.

Manchanda R, Sharma MM (2014) Impact of completion design on fracture complexity in horizontal shale wells. *SPE Drilling & Completion,* 29(01):78–87.

Marshall PW (1969) Risk evaluations for offshore structures. *Journal of Structural Division,* 95:2907–2929.

Marshall PW, Bea RG (1976) Failure modes of offshore platforms. *BOSS '76,* University of Trondheim, Norway, Vol. 2, pp. 579–635.

Marthinsen T, Winterstein SR, Ude TC (1992) TLP fatigue due to second-order springing. Probabilistic methods and structural and geotechnical reliability. *Proceedings of Specialty Conference,* Houston, TX, pp. 455–458.

Masciola M, Nahon M (2008) Modeling and simulation of a tension leg platform. *Proceedings of the Eighteenth International Offshore and Polar Engineering Conference,* Vancouver, BC, Canada, July 6–11.

Mayerhofer MJ, Lolon EP, Warpinski NR, Cipolla CL, Walser D, Rightmire CM (2010) What is stimulated reservoir volume? *SPE Production & Operations,* 25(01):89–98.

McKenna JP (2014) Where did the proppant go? In: *SPE/AAPG/SEG unconventional resources technology conference,* August (pp. URTEC-1922843). URTEC.

Mekha BB, Johnson CP, Roesset JM (1996) Implications of tendon modeling on the non-linear response of TLP. *Journal of Structural Engineering,* 122(2):142–149.

Mercier JA (1982) Evolution of tension leg platform technology. *Proceedings of 3rd International Conference on the Behavior of Offshore Structures,* Boston, MIT, USA.

Mercier RS (1997) Mars tension leg platform: Use of scale model testing in the global design. *Proceedings of Offshore Technology Conference, OTC 8354-MS,* Houston, TX, May 5–8.

Meyerhof GG (1976) Concepts of safety in foundation Engineering ashore and offshore. *BOSS '76,* University of Trondheim, Norway, Vol. 1, pp. 900–911.

Michailidou EK, Antomiadis KD, Assael MJ (2012) The 319 major industrial accidents since 1917. *International Review of Chemical Engineering,* 4(6):1755–2035.

Michel WH (1999) Sea spectrum revisited. *Marine Technology,* 36(4):211–227.

Miller C, Waters G, Rylander E (2011) Evaluation of production log data from horizontal wells drilled in organic shales. In: *SPE unconventional resources conference/gas technology symposium,* June (pp. SPE-144326). Society of Petroleum Engineers, Houston, TX.

Moan TR, Sigbjørnson (1977) Stochastic sea load effect analysis for probabilistic design of fixed offshore platforms. *2nd International Conference on Structural Safety and Reliability,* Munich, Germany, pp. 227–246.

Moe G, Verley RLP (1980) Hydrodynamic damping of offshore structures in wave and currents. *Offshore Technology Conference, 12th Annual OTC,* Houston, TX, pp. 37–44.

Mogridge GR, Jamieson WW (1975) Wave forces on a circular caisson: Theory and experiment. *Canadian Journal of Civil Engineering,* 2:540–548.

Moharrami M, Tootkaboni M (2014) Reducing response of offshore platforms to wave loads using hydrodynamic buoyant mass dampers. *Engineering Structures,* 81:162–174.

Montasir OAA, Kurian VJ (2011) Effect of slowly varying drift forces on the motion characteristics of truss spar platforms. *Ocean Engineering,* 38:1417–1429.

Morison JR (1953) The force distribution exerted by surface waves on piles. *ASTIA File Copy, University of California, Institute of Engineering Research,* Berkeley, CA. AD No. 654.

Moses F (1977) Safety and reliability of offshore structures. *International Research Seminar on Safety of Structures under Dynamic Loading,* Trondheim, Norway.

Moses JF, Stevenson J (1970) Reliability-based structural design. *Journal of Structural Division,* 96:221–244.

Muhuri PK, Gupta AS (1983) Stochastic stability of tethered buoyant platforms. *Ocean Engineering,* 10(6):471–479.

Munkejord T (1996) The Heidrun TLP and concept development for deep water. *Proceedings of the ISOPE,* Los Angeles, CA, May, pp. 1–11.

Murawski SA, Ainsworth CH, Gilbert S, Hollander DJ, Paris CB, Schlüter M, Wetzel DL (2020) *Deep oil spills: Facts, Fate and Effects,* Springer Cham, Springer Nature Switzerland AG, p. 611, ISBN: 978-3-030-11604-0.

Muren J, Flugstad P, Greiner B, D'Souza R, Solberg IC (1996) The 3-column TLP-A cost-efficient deepwater production and drilling platform. *Proceedings of the Offshore Technology Conference, OTC 8045,* Houston, TX, May 6–9.

Murray JJ, Mercier RS (1996) Model tests on a tension leg platform using truncated tendons. *Workshop on Model Testing of Deep Sea Offshore Structures,* IITC, pp. 162–168.

Mutlu Sumer B, Freddie J (2003) *Hydrodynamics around the cylindrical,* revised edition. World Scientific Publishing Co. Pte. Ltd., Singapore.

Ney R, Andrade RFM, Batista RC (1992) Dynamic response analysis of small-scale model tension leg platform. *Marine Structures,* 5:491–513.

Niedzwecki JM, Huston JR (1992) Wave interaction with tension leg platforms. *Ocean Engineering,* 19(1):21–37. doi:10.1016/0029-8018(92)90045-6

Niedzwecki JM, van de Lindt JW, Gage JH, Teigen PS (2000) Design estimates of surface wave interaction with compliant deepwater platforms. *Ocean Engineering,* 27:867–888.

Nivolianitou Z, Konstandinidou M, Michalis C (2006) Statistical analysis of major accidents in petrochemical industry notified to the major accident reporting system (MARS). *Journal of Hazardous Materials,* 136(1):1–7.

Nordgren RP (1987) Analysis of high-frequency vibration of tension leg platforms. *Journal of Offshore Mechanics and Arctic Engineering,* 109:119–125.

Nordic Committee for Building Regulations (NKB) (1977) Proposal for Safety Codes. TS42, September.

Norwegian Petroleum Directorate (NPD) (1985) *Regulation for structural design of load-bearing structures intended for exploitation of petroleum resources.* NPD, Oslo, Norway.

OCS (1980) *Requirements for verifying the structural integrity of OCS platforms.* United States Geologic Survey, National Centre, Reston, VA.

OGP Risk Assessment Data Directory (2010) *Report no.434-1.* Process Release Frequencies, March.

OISD - GDN - 169 (2011) *OISD guidelines on small LPG bottling plants (design and fire protection facilities).* Oil Industry Safety Directorate, Amended edition.

OISD Standard - 116 (2002) *Fire protection facilities for petroleum refineries and oil/gas processing plants.* Oil Industry Safety Directorate, Amended edition.

OISD Standard - 144 (2005) *Liquefied Petroleum Gas (LPG) installations.* Oil Industry Safety Directorate, 2nd edition.

OISD Standard - 150 (2013) *Design and safety requirements for liquefied petroleum gas mounded storage facility.* Oil Industry Safety Directorate.

Onajite E (2013) *Seismic Data Analysis Techniques in Hydrocarbon Exploration.* Elsevier, NY, USA. ISBN: 9780124200234.

Ostgaard K, Jensen A (1983) Preparation of aqueous petroleum solution for toxicity testing. *Environmental Science and Technology,* 17:548–553.

Papazoglou IA, Bellamy LJ, Aneziris ON, Ale BJM, Post JG, Oh JIH (2003) I-risk: Development of an integrated technical and management risk methodology for chemical installations. *Journal of Loss Prevention in the Process Industries,* 16:575–591.

Pasman HJ, Jung S, Prem K, Rogers WJ, Yang X (2009) Is risk analysis a useful tool for improving process safety. *Journal of Loss Prevention in the Process Industries,* 22:769–777.

Passey QR, Bohacs KM, Esch WL, Klimentidis R, Sinha S (2010) From oil-prone source rock to gas-producing shale reservoir–geologic and petrophysical characterization of unconventional shale-gas reservoirs. In: *SPE international oil and gas conference and exhibition in China,* June (pp. SPE-131350). Society of Petroleum Engineers, Houston, TX.

Patel MH (1989) *Dynamics of offshore structures*. Butterworth, London.

Patel MH, Lynch EJ (1983) Coupled dynamics of tensioned buoyant platforms and mooring tethers. *Engineering Structures,* 5:299–308.

Patel MH, Park HI (1991) Dynamics of tension leg platform tethers at low tension. Part I—Mathieu stability at large parameters. *Marine Structures*, 4(3):257–273.

Patel MH, Park HI (1995) Combined axial and lateral responses of tensioned buoyant platform tethers. *Engineering Structures,* 17(10):687–695.

Patel MH, Witz JA (1991) *Compliant offshore structures*. Butterworth-Heinemann Ltd., Oxford.

Patin S (1999) *Environmental impact of the offshore oil and gas industry*. Eco Monitor Publishing, East Northport, 425 pp.

Perrettand GR, Webb RM (1980) Tethered buoyant platform production system. *12th Annual Offshore Technology Conference, OTC 3881,* Houston, TX, May 5–8, pp. 261–274.

Perryman SR, Horton EE, Halkyard JE (1995) Tension buoyant tower for small fields in deep waters. *Offshore Technology Conference, OTC 7805,* Houston, TX, May 1–4, pp. 13–22.

Pilisi N, Wei Y, Holditch SA (2010) Selecting drilling technologies and methods for tight gas sand reservoirs. In: *SPE/IADC drilling conference and exhibition*, February (pp. SPE-128191). Society of Petroleum Engineers, Houston, TX.

Pilotto BMR (2003) Dynamic behavior of minimum platforms under random loads. *Proceedings of International Conference Offshore Mechanics and Arctic Engineering*, Vol. 1, Cancun, Mexico, pp. 441–449.

Pilotto BMR, Stocker R (2002) Dynamic response of shallow water mono-pod platforms. *Proceedings of International Conference Offshore Mechanics and Arctic Engineering,* Oslo, Norway, Vol. 1, pp. 113–120.

Pilotto BMR, Stocker R (2003) Nonlinear dynamic analysis with deterministic and random seas. *Oceans Conf Record (IEEE)*, 5:2908–2915.

Planas-Cuchi E, Salla JM, Casal J (2004) Calculating overpressure from BLEVE explosions. *Journal of Loss Prevention in the Process Industries,* 17:431–436.

Pontiggia M, Landucci G, Busini V, Derudi M, Alba M, Scaioni M, Bonvicini S, Cozzani V, Rota R (2011) CFD model simulation of LPG dispersion in urban areas. *Atmospheric Environment*, 45(24):3913–3923.

Prem KP, Ng D, Mannan MS (2010) Harnessing database resources for understanding the profile of chemical process industry incidents. *Journal of Loss Prevention in the Process Industries,* 23(4):549–560.

Rabia H (1985) *Oil well drilling engineering: Principles and practice*. Graham & Trotman. Inc., London, UK.

Ramamurthy K (2011) *Explosions and explosion safety*. Tata McGraw Hill, New Delhi, India, pp. 288.

Ran Z, Kim MH, Niedzwecki JM, Johnson RP (1994) Response of a spar platform in random waves and currents (Experiments versus Theory). *Proceedings of the International Offshore and Polar Engineering Conference,* Osaka, Japan, April 10–15.

Rana R, Soong TT (1998) Parametric study and simplified design of tuned mass dampers. *Engineering Structures,* 20(3):193–204.

Reddy DV, Arockiasamy M (1991) *Offshore Structures,* Vol. I. Kriger Publishing Co., Malabar, FL.

Rho JB, Choi HS, Lee WC, Shin HS, Park IK (2002) Heave and pitch motion of a spar platform with damping plate. *Proceedings of the 12th International Offshore and Polar Engineering Conference,* Vol. 1, Kitakyshu, pp. 198–201.

Rho JB, Choi HS, Lee WC, Shin HS, Park IK (2003) An experimental study for mooring effects on the stability of spar platform. *Proceedings of the 13th International Offshore and Polar Engineering Conference,* Vol. 1, Honolulu, HI, pp. 285–288.

Rijken OR, Niedzwecki JM (1991) *A knowledge base approach to the design of tension leg platform.* Center for Offshore Technology, Offshore Technology Research Center, Texas, pp. 24–100.

Robert WC, Capanoglu CC (1995) Buoyant leg structure for the development of marginal fields in deep water. *Proceedings of International Offshore and Polar Engineering Conference,* The Hague, The Netherlands, June 11–16, pp. 163.

Rodante TV (2004) Analysis of an LPG explosion and fire. *Process Safety Progress,* 22:174–181.

Roitman N, Andrade RFM, Batista RC (1992) Dynamic response analysis of small-scale model tension leg platform. *Marine Structures,* 5:491–513.

Røren EMQ, Furnes O (1976) State of the art: Behaviour of structures and structural design. BOSS '76, University of Trondheim, Norway, Vol. 1, pp. 70–111.

Sadehi K (2007) Offshore and petroleum platforms for Cyprus oil/gas fields. *Journal of Social Applied Science* 2(4):1–16

Sanderson TJO (1988) *Ice mechanics: Risks to offshore structures,* 253 pp. Graham & Trotman, Boston, MA.

Sarpkaya T (1978) Fluid forces on oscillating cylinders. *Journal of the Waterway, Port, Coastal and Ocean Division,* 104:275–290.

Sarpkaya T, Isaacson M (1981) *Mechanics of wave forces on offshore structures.* Van Nostrand Reinhold, New York.

Sayers CM, Calvez JL (2010) Characterization of microseismic data in gas shales using the radius of gyration tensor. In: *SEG technical program expanded abstracts 2010* (pp. 2080–2084). Society of Exploration Geophysicists, Houston, TX.

Scheidegger AE (1963) *Principles of geodynamics.* Academic Press, New York, p. 362.

Schwartz ML (2005) *Encyclopedia of coastal science.* Springer, The Netherlands.

Sellers LL, Niedzwecki JM (1992) Response characteristics of multi-articulated offshore towers. *Ocean Engineering,* 19(1):1–20.

Shaver CB, Capanoglu CC, Serrahn CS (2001) Buoyant leg structure preliminary design, constructed cost and model test results. *Proceedings of 11th International Offshore and Polar Engineering Conference,* Stavanger, Norway, June 17–22, pp. 432–439.

Sheng JJ (Ed.) (2013) Enhanced oil recovery field case studies. *Journal of the Waterway, Port, Coastal and Ocean Division.* Gulf Professional Publishng, TX, USA, p. 712. ISBN: 9780123865458.

Simiu E, Leigh SD (1984) Turbulent wind and tension leg platform surge. *Journal of Structural Engineering,* 110(4):785–802.

Simos AN, Pesce CP (1997) Mathieu stability in the dynamics of TLP tether considering variable tension along the length. *Transactions on the Built Environment,* 29:175–186.

Sivakumar P, Krishna S, Hari S, Vij RK (2020) Electromagnetic heating, an eco-friendly method to enhance heavy oil production: A review of recent advancements. *Environmental Technology & Innovation,* 20:101100.

Skelton B (1997) *Process safety analysis.* Gulf Publishing Company, Houston, 210 pp.

Skjelbreia L, Hendrickson J (1961) Fifth-order gravity wave theory. *Proceedings of the 7th Conference on Coastal Engineering,* The Hague, the Netherlands, pp. 184–196.

Soding H, Blok JJ, Chen HH, Hagiwara K, Isaacson M, Jankowski J, Jefferys ER, Mathisen J, Rask I, Richer JP, Romeling JU, Varsta P (1990) Environmental forces on offshore structures: A state-of-art review. *Marine Structures,* 3:59–81.

Sondergeld CH, Newsham KE, Comisky JT, Rice MC, Rai CS (2010) Petrophysical considerations in evaluating and producing shale gas resources. In: *SPE unconventional resources conference/gas technology symposium,* February (pp. SPE-131768). Society of Petroleum Engineers, Houston, TX.

Spanos PD, Agarwal VK (1984) Response of a simple tension leg platform model to wave forces calculated at the displaced position. *Journal of Energy Resources Technology,* 106(4):437–443.

Stansberg CT, Karlsen SI, Ward EG, Wichers JEW, Irani MB (2004) Model testing for ultra-deep waters. *Offshore Technology Conference, OTC 16587,* Houston, TX, pp. 1–9.

Stansberg CT, Ormberg H, Oritsland O (2002) Challenges in deep water experiments: A hybrid approach. *Journal of Offshore Mechanics and Arctic Engineering,* 124:90–96.

Stoke's GG (1880) On the theory of oscillatory waves. *Mathematical and Physical Papers,* 1:225–228.

Sun LM, Fujino Y, Koga K (1995) A model of tuned liquid damper for suppressing pitching motions of structures. *Earthquake Engineering and Structural Dynamics,* 24(5):625–636.

Sutherland VJ, Cooper CL (1991) *Stress and accidents in offshore, oil and gas industries.* Gulf Publishing Co., Houston, pp. 227.

Tabeshpour MR (2013) Conceptual discussion on free vibration analysis of tension leg platforms. *Development and Applications of Oceanic Engineering,* 2(2):45–52.

Tabeshpour MR, Golafshani AA, Seif MS (2006) Comprehensive study on the results of tension leg platform responses in random sea. *Journal of Zhejiang University Science-A,* 7(8):1305–1317.

Taflanidis AA, Angelides DC, Scruggs JT (2009) Simulation-based robust design of mass dampers for response mitigation of tension leg platforms. *Engineering Structures,* 31(4):847–857. doi:10.1016/j.engstruct.2008.11.014

Taflanidis AA, Scruggs JT, Angelides DC (2008) Robust design optimization of mass dampers for control of tension leg platforms. *Proceedings of the Eighteenth International Offshore and Polar Engineering Conference, ISOPE-I-08-326,* Vancouver, BC, Canada, July 6–11.

Tait MJ, Isyumov N, El Damatty A (2008) Performance of tuned liquid dampers. *Journal of Engineering Mechanics,* 134(5):417–427.

TNO (1999) *Guidelines for quantitative risk analysis, the director general of labour.* The Hague, The Netherlands.

Vanem E, Antao P, Østivik I, de Comas FDC (2008) Analyzing the risk of LNG carrier operations. *Reliability Engineering and System Safety,* 93:1328–1344.

Vannucci P (1996) Simplified optimal design of tension leg platform TLP. *Structural Optimization,* 12:265–268.

Venkata Kiran G (2011) *QRA in oil & gas industries using PHAST RISK.* M.Tech (Petroleum Engg) dissertation submitted to IIT Madras.

Vickery PJ (1990) Wind and wave loads on a tension leg platform: Theory and experiment. *Journal of Wind Engineering and Industrial Aerodynamics,* 36:905–914.

Vickery PJ (1995) Wind-induced response of tension leg platform: Theory and experiment. *Journal of Structural Engineering,* 121(4):651–663.

Viet LD, Nghi NB (2014) On a nonlinear single-mass two-frequency pendulum tuned mass damper to reduce horizontal vibration. *Engineering Structures,* 81:175–180.

Vinnem JE (2007) Offshore risk assessment: Principles, modelling and applications of QRA studies. Springer London, e-book. ISBN: 978-1-84628-717-6.

Wang KH, Ren X (1993) Water waves on flexible and porous breakwaters. *Journal of Engineering Mechanics,* 119(5):1025–1047.

Wang K-H, Ren X (1994) Wave interaction with a concentric porous cylinder system. *Ocean Engineering,* 21(4):343–360.

Wang Y, Fu H, Liang T, Wang X, Liu Y, Peng Y, . . . Tian Z (2015) Large-scale physical simulation experiment research for hydraulic fracturing in shale. In: *SPE middle east oil and gas show and conference,* March (pp. SPE-172631). Society of Petroleum Engineers, Houston, TX.

Warpinski NR, Mayerhofer MJ, Davis EJ, Holley EH (2014) Integrating fracture diagnostics for improved microseismic interpretation and simulation modeling. In: *SPE/AAPG/SEG unconventional resources technology conference,* August (pp. URTEC-1917906). Society of Petroleum Engineers, Houston, TX.

Waters G, Dean B, Downie R, Kerrihard K, Austbo L, McPherson B (2009) Simultaneous hydraulic fracturing of adjacent horizontal wells in the Woodford Shale. In: *SPE hydraulic fracturing technology conference and exhibition*, January (pp. SPE-119635). Society of Petroleum Engineers, Houston, TX.

Webber DM, Jones SJ, Tickle GA, Wren T (1992) *A model of a dispersing gas cloud, and the computer implementation. I: Near instantaneous release, II: Steady continuous releases.* UKAEA Reports SRD/HSE R586 (for part I) and R. 587 (for part II).

Wilson JF (1984) *Dynamics of offshore structures.* Wiley Inter Science Publications, New York.

Wiltox HWM (2001) *Unified dispersion model (UDM), theory manual,* Det Norske Veritas, Houston, TX.

Winterstein SR (1988) Nonlinear vibration models for extremes and fatigue. *Journal of Engineering Mechanics ASCE,* 114(10):1772–1790.

Yan F-S, Zhang D-G, Sun L-P, Dai Y-S (2009) Stress verification of a TLP under extreme wave environment. *Journal of Marine Science and Application,* 8:132–136.

Yashima N (1976) Experimental and theoretical studies of a tension leg platform in deep Water. *Offshore Technology Conference, OTC 2690-MS,* Houston, TX, May 3–6, pp. 849–856.

Yoneya T, Yoshida K (1982) The dynamics of tension leg platforms in waves. *Journal of Energy Resources Technology,* 104:20–28.

Yoshida K, Ozaki M, Oka N (1984) Structural response analysis of tension leg platforms. *Journal of Energy Resources Technology,* 106:10–17.

Younis BA, Teigen P, Przulj VP (2001) Estimating the hydrodynamic forces on a mini TLP with computational fluid dynamics and design-code techniques. *Ocean Engineering,* 28:585–602.

Zeng X, Liu J, Liu Y, Wu Y (2007) Parametric studies of tension leg platform with large amplitude motions. *Proceedings of 17th International Offshore and Polar Engineering Conference,* Lisbon, Portugal, pp. 202–209.

Zhang F, Yang JM, Li RP, Gang C (2007) Numerical investigation on the hydrodynamic performances of a new Spar concept. *Science Direct Journal of Hydrodynamics,* 19(4):473–481.

Zhang J, Wieseneck J (2011) Challenges and surprises of abnormal pore pressures in shale gas formations. In: *SPE annual technical conference and exhibition*, October (pp. SPE-145964). Society of Petroleum Engineers, Houston, TX.

Zhang QX, Dong L (2013) Thermal radiation and impact assessment of the LNG BLEVE fireball. *Procedia Engineering,* 52:602–606.

Zhang Z, Aktan AE (1995) The damage indices for constructed facilities. *Proceedings of the 13th Modal Analysis Conference,* Nashville, TN, February 13–16, pp. 1520–1529.

Index

Note: Page numbers in **bold** and *italics* refer to tables and figures, respectively.

For Product Safety Concerns and Information please contact our
EU representative GPSR@taylorandfrancis.com Taylor & Francis
Verlag GmbH, Kaufingerstraße 24, 80331 München, Germany